Praise for *Song of Incre...*

W9-CLQ-881

"I have been deeply moved by this book. Jacqueline's work with bees is truly extraordinary. This book has the potential to transform beekeeping and transform the world."

SUSAN CHERNAK MCELROY, author of *Animals as Teachers and Healers*

"Jacqueline takes you on a beekeeping journey to new (or perhaps ancient) levels of interconnectedness. This book is a stunning revelation of deep spirituality that can be found in the simple act of caring for bees."

MARJORY WILDCRAFT, founder, The Grow Network

"It is time for the voice of the bee to be heard . . . Jacqueline's conversation with the bee is a unique contribution to the growing movement of bee-centered beekeeping."

GARETH JOHN, Trustee, Natural Beekeeping Trust

"Whether or not you ever keep bees, this amazingly profound and fascinating book is a deep lesson in the intelligence, wisdom, harmony, and robust vitality of Creation."

BROOKE MEDICINE EAGLE, author of *Buffalo Woman Comes Singing* and *The Last Ghost Dance*

"Jacqueline Freeman weaves two stories. One is hers, a beekeeper who, through attention, stumbles onto an open channel of communication with the bees, and the other is the wisdom of the bees in their own voice. I feel like I can now cock my ear and consult the bees, and that they and others can guide us home."

VICKI ROBIN, author of *Your Money or Your Life* and *Blessing the Hands that Feed Us*

"Jacqueline is on the cutting edge of apiculture with her intuitive, compassionate, and intelligent work. She has the rare ability (and courage!) to take us into the bee's world. We can learn so much from this deeply sensitive book."

LAURA BEE FERGUSON, director, College of the
Melissae Center for Sacred Beekeeping

"*Song of Increase* blends spiritual, practical, and scientific insights into a beautiful anthem for the honeybee."

TAGGART SIEGEL, documentary filmmaker of *Queen of the Sun*

"*Song of Increase* takes you into realms not found in traditional bee-keeping books—realms that require an open mind and also an open heart. The book encourages you to look with new eyes, listen with new ears, and develop the understanding of honeybees required for having the right to say 'I love them.'"

GUNTHER HAUK, author of *Toward Saving the Honeybee*,
cofounder, Spikenard Farm Honeybee Sanctuary

"It is rare to find someone acutely tuned in to the natural world and the many beings that inhabit this web of life. Jacqueline Freeman is one of those special persons. While I was reading this very accessible book, a whole other level of depth perception unfolded as the plant/pollinator relationship took on new meaning. Everyone needs to read this book to understand why it is so important for the bees to not only survive, but thrive."

PAM MONTGOMERY, author, *Plant Spirit Healing* and *Partner Earth*

"This book is a jewel. It speaks to the relationship with the sacred that is so precious. The book is divinely timed, eloquent, honest, compas-sionate, and completely full of love."

DEBRA ROBERTS, master beekeeper, natural beekeeping
educator, Center for Honeybee Research

"Jacqueline Freeman presents a biodynamic philosophy of beekeeping where the science of the sun, the plant, and the soil is woven together with life force, form, and unseen realms of nature . . . Many answers to a healthy future for our bees are to be found in her writings."

CHRISTY KORROW, editor, *LILIPOH* magazine

"If you ever wondered what dreams bees have, what secret role the drones may play, or what the queens sing about, you will discover here a unique perspective—as well as a heartfelt plea to adopt bee-friendly ways in beekeeping and agriculture at large."

DR. LEO SHARASHKIN, editor, *Keeping Bees With a Smile*

"Jacqueline reveals the wisdom aspects of bees and their emotional integrity, and encourages us to venture further into new territory within the multidimensional landscape of honeybees. It is an inspiration for a holistic approach to apiculture and a timely book in a transitional time for bees and the entire biosphere."

MICHAEL JOSHIN THIELE, founder, Gaia Bees

"Beekeeping is a sacred occupation. This book grounds that intuitive knowing and informs my work with my bees. Quite literally, this book has changed my life."

PATTI PITCHER, biodynamic farmer and educator, Farm Wife Mystery School

"This book is a treasure of honeybee philosophy. Jacqueline profiles the many ways bees connect within the hive, with the world outside the hive, with agriculture, and with the beekeeper. Spend time watching your hives. What the bees say in this book is right."

NATHAN RAUSCH, Snoqualmie Valley Beekeepers

"This is the most apicentric book on natural beekeeping that I have read. . . . Throughout the book she lets the bees speak for themselves as they tell us about things hitherto hidden from us, things which could provide a source of inspiration for research far into the future."

DR. DAVID HEAF, *The Bee-Friendly Beekeeper* and *Natural Beekeeping with the Warre Hive*

Song of Increase

Song of Increase

Listening to the
Wisdom of Honeybees
for Kinder Beekeeping
and a Better World

JACQUELINE FREEMAN

sounds true
BOULDER, COLORADO

Sounds True
Boulder, CO 80306

© 2016 by Jacqueline Freeman

SOUNDS TRUE is a trademark of Sounds True, Inc.
All rights reserved. No part of this book may be used or reproduced in any manner
without written permission from the author and publisher.

This work is solely for personal growth and education. It should not be treated as a
substitute for professional assistance, therapeutic activities such as psychotherapy
or counseling, or medical advice. In the event of physical or mental distress, please
consult with appropriate health professionals. The application of protocols and
information in this book is the choice of each reader, who assumes full responsibility for his or her understandings, interpretations, and results. The author and
publisher assume no responsibility for the actions or choices of any reader.

Published 2016

Cover design by Rachael Murray
Book design by Beth Skelley

Printed in Canada

Library of Congress Cataloging-in-Publication Data

Names: Freeman, Jacqueline, author.
Title: Song of increase : listening to the wisdom of honeybees for kinder
   beekeeping and a better world / Jacqueline Freeman.
Description: Boulder, CO : Sounds True, 2016. | Originally published:
   Battleground, WA : Friendly Haven Rise Press, 2014.
Identifiers: LCCN 2015051151 (print) | LCCN 2016026778 (ebook) |
   ISBN 9781622037445 (pbk.) | ISBN 9781622037452 (ebook)
Subjects: LCSH: Honeybee—Behavior. | Bee culture. | Human-animal relationships.
Classification: LCC QL568.A6 F565 2016 (print) | LCC QL568.A6 (ebook) |
   DDC 595.79/9—dc23
LC record available at https://lccn.loc.gov/2015051151

10 9 8 7 6 5 4 3 2 1

To Joseph, for his ongoing love for, belief in, and support of all my ventures. I thank God every day for bringing us to this marriage and partnering you and me in this sovereign life. Bless you, my beloved husband.

When we step into the world of *Apis mellifera*, we are entering a multidimensional landscape of being. The life gesture of the honeybees is so unique and different from most other life forms that a rational mind alone cannot provide sufficient understanding of its nature. Although they are one of the most studied animals, questions about their lives remain. Rudolf Steiner described the bees in his lectures as a "world-enigma." It points toward an understanding that goes beyond reason, and it is an invitation into another mode of awareness.

MICHAEL JOSHIN THIELE

Contents

Foreword

I met Jacqueline Freeman several years ago, when I attended her workshop on the spiritual life of the honeybee. My only experience with bees up to that point in my life had been a sting on the ear from a honeybee when I was six years old. I swelled up like the Michelin Tire Man and kept a healthy distance from bees after that. But on a chance whim—or perhaps in answer to the dream whisper of that little bee from so long ago—I signed up for Jacqueline's bee workshop on her farm.

It was no ordinary class, and Jacqueline is no ordinary beekeeper.

Jacqueline was teaching the class with fellow beekeeper Michael Joshin Thiele, both members of a small and growing cadre of bee enthusiasts who practice an entirely new kind of beekeeping. Called bee-centric, natural beekeeping, or bee guardianship, these revolutionary beekeeping methods emphasize the needs of the bee over the keeper's desire for honey production.

The high point of the class came when we stood before the entrance of a large and active honeybee hive. We were directed to open our arms and send the bees kindness and goodwill. Within seconds, bees poured from the door of the hive, circling us in gentle humming spirals. I stood silent, fearless, and ecstatic in a swirl of bees that circled my body and all our bodies with curiosity and aerial grace. Their hypnotic, thrumming song vibrated deep in my chest long after the bees had departed back to the privacy of their hive.

When I opened my eyes, I saw that I was not the only one with tears streaming down my face. From that day, I have loved bees. I love them with a child's delight and with a mother's fierce desire to protect her family. The bees are our family. We are bound to them and they to us through the grand and terrible process of domestication. They are part of our history and our lives, and our dependence on them is far greater now than their dependence on us. We are tied to them by necessity, for our food survival. The tie that binds them to us is immense generosity.

Jacqueline Freeman is a biodynamic farmer, beekeeper, and artist. She is also a gifted intuitive, as the women in her family line have been for generations. Jacqueline has been receiving messages and images from bees for several years now, and what the bees have told her is being revealed—bit by tiny bit—in scientific studies on the quantum nature of bees, their medicine, their communication, and their organizational structure as a superorganism. This is pretty heady stuff and unconventional to the point of baffling at times, but what Jacqueline has experienced with her bees feels true to me down to my very bones.

This book is a compilation of information from the bees themselves and from Jacqueline's experience with her many hives. I must tell you, this book is full of the word *love*. Love is, I'll admit, an often-overused word. But should you doubt that this word is the least bit excessive in her use of it, I can assure you it is not. Jacqueline is a woman who lives and breathes this word into each and every relationship on her enchanted farm of bees, blossoms, cows, goats, cats, dog, fish, and chickens.

I know this because I have spent many days and nights there while putting this book together. Jacqueline is ceaselessly attentive and loving to whatever creature—plant or animal—crosses her path. In the evenings I find her gathering up lost bees into screen-topped jars, offering them a dab of honey and a familiar piece of honeycomb to rest on for the night. Writing halts for a moment when her cat Remy requests her attention, which she readily shares. Weeding the garden, she'll stop to sort cow and goat leafy treats from compostable weeds and twigs. When the borage plant falls over in the heat, she sends me running for a stake and tie to keep it lifted to the sun. Each evening before bed, she and Joseph bless each animal and their farm.

This is a woman whose words I believe when she shares the bees' message of unity, food made with prayer, and work done with great love. The bees chose the right medium for their message. Jacqueline—who she is and how she lives—is an inspiration to me. She is a woman of confidence, integrity, and great joy; life meets her there in that place, and magic and enchantment unfold around her. This web of goodness she weaves captures all who share her world, from the smallest plant to the most peripheral of associates. She is most assuredly a woman who walks the talk of unity.

Since that time I was enfolded in the spiral of bees, I've taken many beekeeping classes from Jacqueline and from guest teachers she brings to her farm. Jacqueline and her fellow beekeepers are sparking a revolution in beekeeping around the world, and if the honeybee is to be saved, it is Jacqueline and folks like her who will show us how it will be done.

Sitting near the entrances of my own beehives now, I replay the words, images, and wisdom in these pages over and over again. This is not a book to read once and set aside. Keep it on your night table when you need to know there are forces in nature working diligently and tenderly on your behalf. Reading this precious and enthralling book has made me a more hope-filled person—and a far better friend to my bees. My wish is that you, too, may be blessed with the simple grace of bees as you sit with their words and Jacqueline's.

Susan Chernak McElroy
Camas, Washington

Introduction

When the bees speak, I listen.

Strange as it may sound, I do hear bees talk. In my first few years with bees, questions and novel ideas arose and organized themselves in my thoughts, and I found them interesting and useful. Had this continued, I certainly would have imagined myself as having a good connection with the bees.

Then one day, in a moment of reverie, I received eye-opening information about the role bees play in the world, and a whole new understanding of bees emerged. What they had to say encouraged me to make fundamental changes in how I care for my bees.

One becomes a better beekeeper by letting a bee be a bee. If everything is done in service to that, the life force of the colony grows, and the hive thrives. How can we give bees an environment that allows them to have the most advantageous experience of living in harmony with all their purposes? When we satisfy that question, we will have returned to our sacred partnership with honeybees.

How I Fell in Love with Bees

A friend offered us bees, and it seemed natural to accept, as we did when we were given our first chickens. When our first hives arrived on our farm in early summer 2004, I had no experience with bees. I thought of them as another farm animal—one who gave honey instead of milk or eggs.

Like most people, I was quite fearful about getting stung, so I bought a protective bee suit with hat, veil, jacket, pants, and gloves. The first time I put it on, I wore a long-sleeve shirt and jeans underneath to give me a double layer of protection. I pulled on my knee-high farm boots, and I duct-taped the overalls inside the boots so no bees could burrow inside. I taped the edges of my elbow-length gloves over the long sleeves of the jacket. I put on my bee hat and zipped the bottom of the screened veil into the jacket. I even taped the zipper in case a bee might try to get in that tiny opening at the top. Looking like an astronaut suited up for a moonwalk, I marched out to see the bees.

As I approached the hive, adrenaline-filled and bee-protected, I was surprised the bees didn't fly out in a cartoon-like tornado to attack. Gingerly, I put a chair next to the hive and sat down to watch, ready to bound away at a moment's notice. It was sweltering outside, and inside the bee suit it was even hotter. Curiosity soon overcame my fearfulness as I watched bees go in and out of the hive. I watched until my clothing was sweaty and the heat was unbearable.

The bees paid no attention to me. I spent days next to the entrance, my face inches from where they landed and took off, and never once did a bee make an aggressive move toward me. Occasionally, one would land on me the way a bee lands on a branch or blade of grass, with no concern for me at all.

Despite the discomforting heat, I felt relaxed, curious, and happy in a caring way. I began to wonder if the reason the bees didn't chase me off was because I felt so calm around them. Might they be mirroring how I felt? Or was I mirroring them? Could it be that we were connected? I pondered this unexpected thought, wondering if the bees themselves actually thought we were mirroring each other, if *they* were paying attention to me, noticing who I was around them—and if they were causing me to feel this emerging joy. I was on the verge of revelation: if this were true, it would mean my bees were more than mere insects.

Over time I realized the bees could tell my emotional or energetic state. When I embodied kindness around them, they treated me with the same. A cloud of exuberance surrounded us, as though the bees were templating euphoria into the air.

I want you to know I didn't just tear off my bee suit one day and "become one with the bees." That took years. But eventually I did retire my bee suit. The first time I walked right up to the hives wearing only a T-shirt and shorts, I felt a bit anxious and self-absorbed, but then I remembered to turn my thoughts away from myself, to open myself to the bees and let them feel me out—which they did. They landed on my bare arms and licked my skin for the salty minerals. When I held a finger next to the entrance, a sweet little bee delicately walked onto my fingertip and faced me. She looked right into my eyes, and for the first time, we saw each other.

And so I became part of bee life.

Becoming Kin

I soon found myself having more intuition about the hives. One morning in early spring, before the flowers had come into bloom, I suddenly had the idea that I should check one of my hives. I found the bees unexpectedly out of food; so I fed them honey saved from the year before. That call I intuitively heard from the hive likely saved its life. Another time I had the feeling that a distant hive in the east pasture was on the verge of swarming. When I walked up to see, sure enough, they were. Events like this taught me to trust my intuition more, and listening to my intuition continues to bring me into a closer relationship with all the hives.

In my sixth year with bees, something new happened. I had begun a morning practice of contemplation, quieting my mind and opening my heart. I entered this prayerful state, asking for guidance, direction, courage, and truth. Even though I didn't mention honeybees, they immediately began appearing in my thoughts and passing me information I had never read or learned from other sources. I believe the sincerity of my questions opened a door. When the information began coming to me, I listened with attentiveness, respect, and gratitude. The more I listened, the more information they shared.

Since my first intuitive conversation with the bees, I have had many others. At first I didn't know how to explain where the information came from, and that bothered me. I told my husband's friend, Steve Hall, that saying the information just pops into my head sounds strange, even if it is true. Steve is a holistic physician with a broad knowledge of many different sciences. He told me about numerous occurrences in science when revolutionary ideas emerged through similarly curious and intuitive channels. While not common, information that arrives in this way has a solid history of shifting scientific thought forward.

For example, the German chemist Friedrich Kekulé dreamed he saw a snake seize its own tail. Upon waking, he connected this image to a problem he'd been working on and finally identified the elusive ring shape of the benzene molecule. Linus Pauling published the first paper identifying DNA as housed in the shape of intertwined helixes—information that had come in a flash of inspiration. Pauling's ideas so often came to him this way that he set aside daily time to listen for the knowledge. Pauling received two Nobel Prizes for his discoveries. Nikola Tesla, a brilliant scientist with numerous inventions to his credit, openly said that many of his ideas—including the invention of alternating electrical current—originated from intuitive inspiration.

In a similar way, the bee teachings enter me. For the most part, the information comes to me fully formed, as if I'm reading it out of a book. Occasionally I see images, but mostly I hear full sentences. Sometimes complete essays came forth, needing little more than punctuation. During certain readings I've felt distinct emotional and physical responses, such as when the bees described the quickening during the queen's elation-filled marital flight, the debilitating feeling of sugar in a bee's belly, or the shared jubilation of swarming bees. In each of these experiences, the feeling gave more depth and color to my understanding.

Sometimes what the bees tell me is so far beyond my own intellectual base, I have to stretch to understand their meaning. The bees share information as they feel I can understand it. They often give me a broad overview first; then they explain the underpinnings and details. Sometimes I ask for the answer to a specific question, but they always tell me what I need to know first. When my understanding is sufficient, they address what I initially asked about. For instance, in three consecutive readings, I had asked what they call the worker bees. I thought they had ignored my question. When they finally answered, I realized I had needed to understand how they experience sound before I could rightly hear the vibrational word-song that represents worker bees.

You can label my communications with bees as clairvoyance, higher intelligence, a deeper connection with nature, animal communication, or the ravings of a wild bee fanatic. All I know is that I'm a better friend to bees because of what they've taught me.

The Hive Is a Holy Place

I come to the hive as I would enter a holy place. I come to the bees' presence with reverence, respect, gratitude, and generosity. These are the qualities bees bring to their interactions with each other.

Most people believe the role of the honeybee is to pollinate crops and make honey for humans to eat. Many of the bees kept by humans are pigeonholed into those two roles and are often treated as if indentured servants whose mission is to serve our needs. Conventional beekeeping methods are human-centric, designed to pressure bees to produce more product in less time as they work for us.

Our present attitudes and beekeeping routines have strayed from bee-centric methods and are the root of the problems bees are having these days. These conventional techniques may serve the marketplace, but they aren't always bee friendly, and compromises are often made at the expense of the bees. This narrow thinking has kept us blind to the tremendous depth of knowledge that bees embody and to the generosity they carry forth each day to serve the highest needs of our spiritual development and the evolution of the world. Surely that is a broad statement, but I stand behind it. This requires us to take a second look at how we share this planet with other forms of life.

If we look from the perspective of a bee, we become capable of asking what bees most want and how to care for them in ways that put their needs first. To do so requires an understanding of bees who live with little or no human intervention. If we know what bees do on their own and why they do that, we may be able to provide our honeybees with a similar environment where they can flourish.

Treating bees with reverence and gratitude will do more to help them than you can imagine. As you read about the many profound ways honeybees offer their work to propel humankind's spiritual evolution, I expect you will be inspired by their respectful industry and fellowship. They are living examples of love, of an interdependent community, and of an ever-outflowing story of creation filled with patience, kindness, and compassion. Daily, they live teachings we could benefit from learning.

I have had plenty of opportunities to share this new information with bee folks. For many years I've been teaching bee and agricultural classes at our farm. Every spring I teach people how to work with itinerant swarms and how to care for feral bees in ways that respect their wildness. I've spoken at regional and national conferences, and I appeared as the swarm rescuer in the documentary film *Queen of the Sun*. My work appears alongside that of people I admire, such as Vandana Shiva, Raj Patel, Gunther Hauk, Michael Pollan, and Michael Joshin Thiele. And, through a stroke of what I consider luck, I'm featured in Chris Korrow's documentary *Dancing with Thoreau* along with the Dalai Lama. Each year my husband and I hold a bee conference with other bee-loving folks who are also waking into a new—or perhaps very ancient—and respectful relationship with bees.

About This Book

The book is structured so you can learn about the true nature of bees, understand bee behavior, and develop a consciousness that enriches your interactions with and appreciation of them. I've included many stories that show the way I interact with bees. This ongoing bee-human relationship has taught me to understand how the colony imagines itself, how bees direct their activities to their very special place within Nature, and how they carry the world forward each day. In the sections called "In Our Own Words," I provide contextual details so you can fully understand what the bees say through the words they revealed to me.

In all respects, this book has been co-written. The relationship I have with bees isn't with any one hive or a single bee; it's with all bees. Beekeepers sometimes ask me to talk with one of their hives to find out what the bees need, but that's not what I do. This one-on-one relationship is meant to be between the beekeeper and those hives under the beekeeper's care. When I speak to the bees, I am speaking with a knowledgeable presence that embraces all bees, a consciousness that understands and wants us to know and be respectful of the purposeful actions of the bee kingdom. I've done my best to convey what they've taught me.

The framework for the book is the many "songs" the bees sing as they go about their tasks. They use the word *songs* to refer to the different times and tasks within the hive, their activities, and seasons. For example, the time of increase happens from midspring to early summer, when everything within the hive is abundant, fertile, and growing larger each day. The hive is so successfully expansive that its consciousness knows when it is ready to send half of the hive, along with the current queen, out into the world to become a new colony on its own. It's a time of intensifying excitement. At this joyful time, the bees sing the Song of Increase. It is a celebratory anthem. If you find yourself standing next to a hive as they sing the Song of Increase, you'll feel that song in your bones, an exhilarating upsurge of multiplying joy. Ah!

I call myself a "relational beekeeper." But this book is not about beekeeping methods, treatments, or systems. Instead, the only beekeeping "method" I offer to you is one of kind observation, creating supportive homes for bees and fields for them to live in, and tending the heartfelt relationships that form when we are with them. My hope is that your relationship with bees—as a keeper, a gardener, or simply a caring friend—becomes gloriously rewarding for you and the bees.

As my husband, Joseph, is fond of saying when he hears these stories, "They have the ring of truth to them." I invite you to listen for that ring of truth as I share these insights from the bees. What I've learned at their wings is the most profound education, and I am grateful to the bees for sharing their world with me.

I

The Song of Unity

*How Bees See Themselves,
Their Colony, and the World*

We begin our journey into the world of the honeybees by exploring their largest self first: the colony, the hive. From this vision of their wholeness, we first glimpse the elusive, mysterious wonder of unity. The honeybee is an exemplar of unity consciousness. In fact, the term the bees use to speak of their colony is *the Unity.*

The Unity can also refer to the environs the colony is a part of. As part of their world, we, too, are part of that Unity. When the bees speak about embracing us in the Unity—a theme they return to often—I think of the many times I've sat beside my hives and been privy to exactly that. Though I am one of many, I am accepted into the embrace, and I feel myself part of the entire and blessed by nature.

The idea of being in unity with all that surrounds us at first surprised me. I was used to having personal territory that I assumed needed protecting. Being in a place of unity, wherein I am enfolded into a shared life, is curiously different. I feel both supported and protected, which in turn makes me feel more open. As I sit next to a hive, watching bees flying in and out, my mind clears of conscious thought, and I enter a deep state of bee meditation. Many bee stewards, I'm sure, know this state. We sit next to the hive, and as the sound of the hive enters us, we find ourselves in a deep reverie that opens our hearts. The bees draw us into our opening heart and welcome us there. It is a heart filled with great love and great activity.

IN OUR OWN WORDS

We wake up to the understanding that we are all one, all the time. Human beings exist connected each to each, but believe that they are not. Honeybees dwell in the full realization of that connection and have done so for eons. The unity we embody is a reflection of the kingdom-wide Unity that dwells in us all.

This is the gift we bring: complete, sacred unity in body and spirit. To be in the presence of Spirit [God], to simply sit and *be* in such presence, offers the opportunity to be transformed by it. This we offer you. Come sit. Be with us. Drink in the Unity as you would fresh rain. We offer our gift with great joy and love!

Wholeness: Embedded in the Task

Bees live in wholeness. They dedicate themselves to working within the hive in a way we humans don't grasp because we have individual personalities. Bees have the hive, and each individual pours love into the care and keeping of the colony.

A typical bee colony is made up of thirty to fifty thousand honeybees living inside a magical, womblike enclosure. Though individuals, every bee within the colony works for the good of the hive to help it function perfectly. Honeybees are interdependent, relying upon each other to create the working hive that allows and encourages them to flourish. They live in a perfected communion most of us lost long ago. The bees help jog our memories of what it is to live a life of devotion, joy, and loving membership in a strong, committed tribe.

Each member of the hive community dedicates 100 percent effort to all interactions. That's quite an interesting idea to me because I was raised with the idea that fifty-fifty is the goal for a two-person relationship and that, in larger human groups, each person giving a little bit is enough to carry a group forward. The hive model of 100 percent effort given to all tasks by every bee is a wild idea that got me thinking: What if we humans set aside our indomitable nature and put the betterment of all in front? What would that do to our relationships, communities, and the world?

Intrigued by such a brave and generous commitment, Joseph and I decided to give this crazy idea a try in our own household. We've been married a few decades, and as all couples do, we have devoted plenty of time solidifying the cause of our conflicts. An example would be who does the dishes. While we weren't exactly keeping score, if one of us did dishes three times in a row, it certainly would be brought up as a reason for not having to do dishes that fourth time. Being willing to do everything 100 percent instilled a sense of order we hadn't expected. Now, if I do dishes six times in a row—without complaint—chances are my husband has also already stepped up to doing more of the things I don't relish, like making dump runs or keeping our vehicle tanks topped off so I don't get my hands dirty pumping gas. Truthfully, I don't mind doing dishes much anyway, and he doesn't mind digging deep holes to plant trees (something for which I lack talent). While we don't pretend perfection reigns in our marriage, we are pleased with how this idea has changed our behavior. It keeps our marriage focused on true partnership.

Every bee is totally committed to doing whatever the colony requires. When a need comes up, the task is answered without hesitation. I've seen a bee signal to another, "I've got an itch I can't reach" or "The floor is sticky" or "I dropped a pollen pellet," and other bees jump in to help. Once, while working inside the hive, I inadvertently broke a honeycomb, which fell on the hive floor and made a mess. A whole posse of bees quickly started licking up the honey and carrying it off to store. No one said, "I'm busy." Everyone jumped in and cleaned up the mess and then went back to their other tasks. I could tell by their hum there was no blame, frustration, or anxiety, just easygoing cooperation. Each bee does what needs to be done to move the colony forward. How utterly divine that they put the colony first.

Imagine if we did that. Imagine if each day we put our best self forward and did whatever it is our community—local or global—requires to keep the world going in a way that supports all life. Could we be so brave and generous?

The Bien: Living in Unity

Science describes a honeybee hive as a superorganism—a single entity made up of tens of thousands of individuals working together and functioning as one living being. Thus, a honeybee is both an individual being and a cell in a larger being. Each bee is both particle and wave in the physics of its world.

Some well-attuned apiculturists, such as Michael Joshin Thiele and Gunther Hauk, refer to the hive community with a more comprehensive German word—the *Bien*. Besides having the qualities of the superorganism, the Bien includes the spiritual center and life force of the hive. Collectively, the Bien is conscious and alive, like a living brain, operating as a unified thought. The Bien is not just the hive; it is this hive in this particular and sacred place. The Bien also encompasses a relationship to place, light, seasons, and plants.

Within this holy landscape of the Bien, bees work with flowers, trees, minerals, water, light—the full spectrum of Creation—to bring about the highest good for our earthly environment. Unknown to most humans, bees also work intimately with the unseen realms of nature spirits, elementals, and faeries. All actions are both for the good of their singular Bien and for the unified consciousness of all Biens everywhere.

A single honeybee alone can't perform the many necessary survival tasks a bee needs to do, nor can a honeybee survive without its hive. If a foraging bee gets lost or trapped during her travel and is stuck somewhere overnight, most likely she will be dead by morning. I believe her death comes from the separation from the Bien.

Each area of the hive and the tasks of all the bees depend upon each other for full expression. Bees understand that they do not exist as solitary beings, though they each have individual roles. Rather than saying they act in unison, like a group that works toward a common goal (which they do), they reference a larger perception of awareness. They are beings who carry memory and meaning through time. While each bee lives a short life, the hive itself continues on and carries the wholeness of its expression through time. They use the word *memorist*, which was new to me the first time I heard it, to describe the hive's collective memory.

Bees progress through a chronological sequence of tasks over their lifetime. As old bees die, new bees step forward to fill the prior bees' roles. They take on the tasks with a fully expressed presence that expands into the history and future of each bee. The consecutive line of bees continues into time, and the hive functions as a perpetual presence. Each hive is continually populated by bees into eternity. Although individual bees die, the Bien is capable of living forever. In their world beyond the hive walls, bees bring life, song, and spirit to all they touch, making their larger community a more vibrant, abundant place.

We humans have tasks and are aware of our personal history, but most of us feel somewhat disconnected from our human history. We know stories from the past, but we don't readily embody the lessons and knowledge of our history, and we certainly don't grasp our future. Some humans, like indigenous Aborigine tribes of Australia and Pygmy tribes of Africa, hold the tribe's memory and, indeed, even its future awareness in their ken. They are connected to each other through a shared history and through knowledge of a shared future—a way most of us have forgotten or cannot imagine.

Sitting quietly beside a beehive—or sitting quietly in any natural setting—is one simple way of finding our way back to such a connection. When we quiet our minds, we rediscover our goodness. When we embrace our place within the wholeness of our landscape and our community, we open to our larger, better selves. During reflective moments with my bees, I hear something indefinably enriching and fulfilling in the hive's sound—what they call a "single tone made up of thousands"—and I am bettered by it.

Imagine if we woke each day asking ourselves: How could we better the world? How could we sing ourselves into aliveness? Who would we be if we understood that we are all connected, that our individual efforts have great value when we undertake actions that better the world whole?

IN OUR OWN WORDS

A hive is a wholeness. One bee experiences all that every bee experiences. There is no separation.

Whereas you have separation, we are born into a world beyond the borders of singularity. Our first thought is always of the hive, to bear increase in the world as we sing the world into aliveness. In a unity of purpose, bees have the fortitude and attention to meet each task with a steadiness of spirit. Each bee does the task at hand, doing what needs to be done to carry the hive into its fullness of being.

We are not soloists, though we each make our own sound. We are memorists who have remarkably retentive memories. We come to the hive, and we are the hive. We sing our Unity and then each take our song out into the world. We touch each flower and deliver the signature of our creation. After our touch, each plant has a rising helix, a chromatic cord, that joins earth, matter, and ether. The fecundity of the atmosphere is thus enhanced and enlivened.

Pollination is much more than fertilization. The act of pollinating moves reproductive forces and, at the same time, enlivens the ether. Pollinating and the daily revitalizing of the ether is our task. We are embedded into the task, as it is in us, and we have no singularity in it.

Bees and humans are different in our forms, yet we are unified in our mission of spiritual evolution. We bring knowledge of these spiritual forces that you may know how love and generosity of spirit create the world anew each day.

Where the song of the hive is strong, the nature spirits flourish. The hive is a beacon of light where work and love go hand in hand. To the nature spirits, each hive and each bee is wrapped in light. The sound of an industrious hive is so full of life that it feeds the soul. Hives are sources of spiritual nourishment for the nature-spirit kingdom and are places of deep reverence. Nature spirits come here to recharge and rejuvenate themselves, filling themselves with spirit. The hive light is living food, living nourishment. Hives are shrines that generate great amounts of healing, loving, and creative energy. Bees and their hives are temples of the spirit world, source points of spiritual manna.

Humans have individual souls; the hive and all hives have a group soul. Animals have a group soul individualized; bees have a group soul intact, never separate. Each bee is part of a fully experiencing group soul. We live and breathe in spirit always, fully connected. In this evolution of the group soul, every hive has consciousness of every other hive. We unify in a harmony, all sounds finding a commonality—a single tone made up of thousands.

Show yourself to us that we may embrace you into our Unity.

Collapsing Colonies:
All Damage Has an Effect

Bees make honey, and even if that were all they did, I'd applaud their efforts. On a larger scale, bees pollinate more than one-third of our crops, and without them, much of the produce in our stores would disappear. But there's more: somehow, magically, their simple presence in a field increases the production of fruits, vegetables, herbs, and flowers significantly—up to 25 percent—while also increasing fruit set and size. (For just one example of how bees benefit crops, you can find a report called "Effects of Honey Bee Pollination in Pumpkin Fruit and Seed Yield" in the April 2006 issue of *Horticultural Science*.) We need bees.

In 2006, one-quarter of the bees in the United States died. Since then, bees by the millions have been abandoning their hives, and hundreds of thousands of bees have been falling dead in cities and across farms and fields. All over the world, honeybees have been vanishing. This catastrophic decline is referred to as *colony collapse disorder*.

No single cause has been found for this decline. Some blame it on parasites, viruses, environmental poisons, or genetic alteration of crops. As of 2015, research heavily implicates a particular brand of pesticides called neonicotinoids. When a friend of mine contracted cancer at a young age, her oncologist told her it was the result of "a thousand little insults, over and over again, to the body and spirit." In the same way, it is likely that colony collapse disorder is the result of many overlapping causes.

When I spoke with the bees about colony collapse, they spoke about relationship. Everything the bees do is about relationship with one another. The story of colony collapse is a story of how these relationships have been broken, contaminated, or subverted. It is a story of ignorance, thoughtlessness, and selfishness—qualities we humans bring to far too many of our relationships, from the most personal and intimate to the most global and institutional. In the chapters to come, you will hear much about the complexities and the subtleties of relationship, because right relationship truly is at the center of Unity consciousness, which is the heart and soul of the Bien.

One specific relationship is critical to bee health and survival: the relationship between bees and pollen. The bees say colony collapse begins with pollen. They have a deep, complex, and ethereal relationship with plant pollen and require a multitude of pollen types to keep the colony in good health. Yet today's agriculture is directed at thousands of acres of a single crop, creating a situation that is harmful to bees. These vast swaths of monoculture offer only minimal varieties of pollen, and much of it is spray-poisoned, gene-manipulated pollen. Suddenly the door is open to thousands of little insults to the body and spirit of the Bien.

Nature has a mission to bring to the future, through offspring, each life form and its genetics. If the life form is weak, the reproductive process is shut down so that deficient genetics don't carry forward into the future. To do otherwise wouldn't make any sense and would weaken the line. Current statistics say that 15 percent of human couples in the United States are infertile. The medical response has not been to strengthen the life force of the infertile couples, creating healthy human beings who can easily carry their babies and their lovely genetics forward. Instead, it's been to find more clever ways to trick human bodies into becoming pregnant. Hence, we have fertility drugs and multiple egg implantations leading to octobirths.

In much the same way, it appears that many honeybee queens are missing the life force required to be viable mothers of future bees. Instead of supporting hive health, we typically overmanage and interfere with colonies in ways that cause harm to the bees. A few examples:

- We force-feed bees sugar and corn syrup instead of allowing them their natural, nutritious honey.

- We scramble the genetics of queens and drones.

- We introduce new colonies of bees to regions where they are not acclimated to the local climate.

- We prevent swarming, which compromises the queen's fertility.

- We create stress by constant in-hive manipulations.

Fixing this missing life force at the source involves treatment-free beekeeping, appropriate food (honey, not sugar), clean forage, and getting away from conventional, factory-farming methods of raising bees. That is a big checklist, but we have to start somewhere if we expect bees to survive in our human-tainted world.

IN OUR OWN WORDS

Although the hive itself is conscious of each undertaking and of the purpose and means of completing each task, we fret when these projects are not accomplished. The colony is aware of its strengths and weaknesses and works to surmount its insufficiencies.

The colony desires that which makes it whole. We try to remedy our problems—and many hives do—but some hives are not able to survive. Though they have life force and they work to fulfill the tasks of a colony, these bees cannot fill the directive to thrive because they are broken. Hives populated with these wounded bees are markers, given the task to evolve or devolve when exposed to things like pollen with an altered gene structure or nectar that can damage organs.

Each bee aligns itself with cosmic forces that direct its role in the colony. The hive's daily functioning is precise and rhythmic, with each bee's task divinely orchestrated to fulfill the role of the hive.

Pollen emits a sound vibration that fits neatly into the historic memory bees share. When pollen is chemically altered, the atomic structure has a strange vibration that doesn't fit what we know as common to our perception and knowledge.

A hive can accommodate some insufficiency within the pollen's vibratory expression, because there is enough volume to make up for a bit of weakness. The wide variety of pollen sources provides ample stores for the hive. But if only one kind of pollen is made available to us, as in monoculture cropping, where bees are trucked to pollinate thousands of acres of only one kind of flower, we cannot gather the wide variety of necessary herbal, tree, and flower pollens necessary to regulate our immune systems and fight off disease.

As we collect and bring home the pollen, its vibration tunes us to a channel that bees work within. Clear, pure pollen, rich with life force, directs us in the proper action and relationships with all that surrounds us. Our sensitive antennae are tuned to a channel that gives us the ability to communicate with other bees, as well as stay connected to our Creator. The more vital the pollen, the more it dials us in. Each bee knows its task and purpose via the constant communication that comes through the clear channels of the antennae.

Pollen is absolutely primary. Pollen is the determinant of the hive's health. The clarity of our actions is determined by the purity of the pollen and the strength of the channel. When pollen is molecularly altered by exposure to poisons, the poison damages the genetic material and its vibratory expression. When pollen is nutritionally weak, it does not provide sufficient energy for the hive to thrive. When only one pollen is available, bees sicken.

The pollen fermentation process can align the chemical structure of damaged pollen to a degree, but when there is too much damage, fermentation can't fix it. All damage has an effect.

When a bee is born, its antennae are attuned to the channel of the bee's highest expression. When a young bee has had insufficient nutrition and inadequate vibration from the pollen, the bee suffers. Its channel isn't tuned in clearly. It lives with a modicum of static that distracts and inhibits its access to full knowledge and expression. A hive altered by impaired pollen bears a dispiriting hollowness within its sound expression, which affects all levels of the hive. While a hive can carry that condition for a while, ultimately, the lack becomes predominant, and the hive is aware that it is coming up short. This static-like sound continually distracts and draws energy from the tasks at hand and inhibits the singular focus of the bees.

Humankind remains unaware of how significant its alterations are to chemical structures and how detrimental these alterations are to finely tuned channels of expression, such as bees. Poison alters the environment on so many levels.

When toxic exposure to living beings is gauged, the results are only measured in human scale; what humankind cannot measure is deemed insignificant and, thus, allowable. Even when toxic damage is acknowledged, it may be permitted nonetheless. No weight is given to how genetic aberrations damage the vibration and inhibit the life force of bees.

Each hive has an Overlighting Being. The Overlighting Being is the representative of the hive and, at the same time, has responsibility to the bee kingdom. The Overlighting Being ensures that each hive is aligned with the highest expression of its bee-ness and that it contributes to the bee kingdom's evolution. The Overlighting Being is the repository of the hive's history and the emissary who speaks on its behalf. When the bees bring in pollen, the Overlighting Being revels in remembrance and appreciation of each plant's genetic material and vibratory expression.

When a hive's attunement to purpose is thwarted by exposure to chemicals, molecular disorder, genetic disruption, or atomic disarray, the effects are heinous and devastating. When a hive loses its signal clarity and gets skewed, it becomes unable to find its center. The richness of sound that invests the hive's core is diffused and dispersed; its connection to purpose is obscured and ultimately lost. At this point, the Overlighting Being, acting in correct relationship with nature, invites the hive to remove itself from the bee family.

The hive recognizes that it cannot stay tuned in to its evolutionary purpose, so the colony removes itself. The bees sacrifice themselves so their weakness doesn't carry forward.

Each hive must have the autonomy of serving its own needs, while also being in service to the larger purpose. The ability for autonomous thought is our protection. Humans, in their folly, are making decisions for bees; but these decisions, without our input, put us at risk.

Monocultures are a betrayal of the trust between bees and humankind. Humankind cannot have monocultures without migratory beekeeping. This false movement of bees must cease, as it is too hard on hives. Hives need to decide, on their own, how to make hives strong.

Let our natural practices express through our own tastes and timing. We ask that humankind enter with us into a revolution of agricultural practices, where together we seek respectful relationship with all beings.

II

The Song of Belonging

The Sacred Work of the Queen,
Drones, Maidens, and Pips

Within the great breadth of the Bien dwell the small bees, like cells in a brain or thoughts in a vast mind. Each individual bee has an important role in the hive's successful function, and we humans have given them names that evoke certain roles in the human community, such as queens or workers. We've named the male bee by his sound: drone. But all of these names minimize their life, work, and roles. A queen is not so much a hive monarch as the hive mother. The word *worker* implies a slight tone of drudgery, while *drone* hints at a sound no one wants to hear, as in "droning on and on."

Within the hive body and out in the fields, the individual bees go about their tasks with astonishing focus, devotion, and energy. The queen pours herself into her work as mother to every bee in the hive. Other than to ask for a drop of food, she does nothing on her own behalf. She has no task other than to serve. The workers—whom the bees call maidens—take up many professions in their short life and do them all with gusto and joy. I know this from speaking with the bees, but any perceptive beekeeper feels the sense of delight and enthusiasm at the door of most hives.

In conventional beekeeping, drones are thought good only for impregnating queens. With the hive's queen already pregnant, drones are commonly labeled useless disease vectors and a waste of resources within the hive. Yet bees themselves call drones "the holiest of beings," who function much as holy shamans within their own and neighboring hives.

Maidens, drones, and queen work together in a cooperative culture that celebrates joy, calling, and beauty.

IN OUR OWN WORDS

We are the measure
Wing to tip, hand to hand,
All the drones, cell by cell,
As throat of flower
To length of tongue.
As time before love's last forage.
Track of sun our distance dance
As egg to pip to cycling ray
As seasons go, the comb now filled
As queen's round heel, forever ever.

The Maidens: Messengers of Light

One day I saw a drone land at the entrance of the hive and begin to wiggle to and fro in place, signaling some kind of upset or concern. When something is not quite right with a bee, like when it feels itchy in an unreachable spot, that bee will squirm around and even brush up against other bees, telling everyone that something is tickling, biting, or bothering it and asking for help to remedy that. This drone was calling for a cleaner bee to help get an itchy thing off him before he went inside the hive.

A moment after his wiggly signal, a maiden dashed over and gave him a thorough cleaning. She climbed on top and scoured his back, reached under his wing surfaces and the wing joint. As she examined his abdomen, she found something and bit it; she then jumped aside and spat it out. I couldn't tell what it was (though I suspected it was a mite) because she lunged on it and bit it again. Then another bee jumped at it, and the little speck fell off the edge of the landing platform. The cleaner bee stepped aside, and the drone walked calmly into the hive.

For every task of the hive, there's a maiden ready to do it. Indeed, the maidens perform most of the activity in the hive, and their cooperation makes a hive fully functional. They are models of harmony, integrity, and devotion to the hive. So strong is their devotion to the hive that if separated from their family overnight, they may perish of loneliness. Author Gunther Hauk, in his book *Toward Saving the Honeybee,* describes the maidens this way:

> Truly, the term "labor of love" would apply to the
> workers, whose selfless activity is a source of marvel and
> amazement. The concept of "love" is not used here in
> the way Hollywood presents it, . . . but rather in its true
> meaning, namely that "love" is work; work utilizing and
> applying wisdom, insight, understanding.

Maidens compose 90 percent or more of the hive's year-round population and nearly 100 percent in winter. After twenty-one days in a honeycomb cell, a maiden bee is born. She chews her way out of the cell and turns nearly at once to assist the nurse bees by cleaning and tending the brood cells in the nursery. A few days later, she becomes a nurse bee who feeds the pips (the bees' word for their babies) her glandular secretions of royal jelly. A tiny bit of royal jelly protects the larvae from bacterial infections and, in more quantity, is the steady diet of any queen cells in gestation.

As the maidens mature, they build comb, feed and care for the queen, maintain a communication network, inoculate new bees with intestinal flora, and feed everyone in the hive. They maintain the ideal temperature to keep the brood warm year-round and bring in water as needed. They gather, process, and store nectar and pollen. They keep the hive's air healthy, regularly clean out dead bees, and clean live bees and the hive itself. They defend the hive from intruders, pests, and diseases; seek out nectar and pollen sources; and carry everything back to the hive. They create propolis and use it to seal the hive. They prepare for swarming and, when necessary, find a new home. If the queen dies, a maiden may step in and lay eggs (more on this later). If any one of these tasks is found wanting, the hive suffers. If all are done well, the hive thrives.

In busy summer months, a maiden bee's life expectancy is about forty-five days. They work themselves to the full expenditure of their capacity and die with tattered wings, after thousands of trips to tens of thousands of flowers.

IN OUR OWN WORDS

We, the maidens, are the messengers of light. Our tasks are more than the work at hand. Though we may look small, each fills our role, and thus bee, hive, kingdom, all phyla, and the world move forward together.

We ask you to bear this knowledge that you may know our shared evolution:

> Little bee with a thousand eyes,
> Tall human with two,
> The rising sun pours into us.
> The light fills us with awe,
> Bare naked, we fear ourselves alone.
> Robed in awareness, we know each matters.
> Each is cousin to the other.
> Illuminate with these words.
> Hear our song. Know our task.
> Share our joy, as each day breaks,
> We create the world again.

Evolution isn't random. We all work together. The act of working together is the evolution. Cooperation among us accelerates development of each species. The way we hold and support ourselves and each other advances our shared evolution.

Everyone has a voice, from the littlest to the biggest. Some know and hear the voice of all that lives. Others say that can't be true because it shakes the foundation of their faith.

We beings of light ask you to know that we are all connected. Would you ignore that and choose to believe some are separate, some are dominant? We are embodied in the knowledge of our connection. Some of you believe you stand alone. In our beautifully connected world, nothing stands alone. There is consciousness in everything. This is God's voice and how God speaks to us.

Naming the Maidens

My friend and fellow beekeeper Michael Joshin Thiele told me he found it hard to believe that the bees would call the hive's females *worker bees,* as humans have named them. Though the females do much of the colony's tangible work, he thought it disrespectful to call them by their job titles and wondered if the colony called them by a more comprehensive name.

When I asked the bees what they called the females, they responded with a three-part lesson, shared with me over a few days.

First they told me more about the interior of the hive, the safety of the area where they begin their lives. Next they described the tasks they learn as they move from one job function to the next within the hive. Then they explained how they generate a sequentially developing sound, much like a Tibetan chant. The sound begins focused on the bee's internal connection with itself and then transforms to an open, expansive tone that flows from the bee and reverberates out into the world. Like an OM in reverse, they said. This development is opposite that of us humans, who at first are external to our own perceptions and then grow and progress to knowing our interior world.

Along with this deep explanation, they also answered my original question: they call the females of the hive maidens.

IN OUR OWN WORDS

The Arc of Creation opens with a blessing, a sound that names us. We dwell in the enclosed area, the interior of the hive, for the first part of our lives. The interior space is a safe enclosure filled with industry.

When we mature, we move outdoors into the larger world. Our name is the interior space that comports us to the exterior, all while in a shared space of consciousness.

As a human, you wake up in an external world, and you work your way to learn of the interior. You know the external world as the open sound of OM. As you progress in your development, you come to know your interior world, and the sound moves into you. Thus, OM begins with the sound of all the world and then carries you inside to the sound of one.

We come into the vast interior world embodied in service, and then we go out into the land-world with our gift. Our name begins in the enclosed space of the hive, in the surrounding *hummmmm,* in and through all of us. When we mature and then transition outside to foraging, our sound opens to the wide world. Though the image implies that time moves from one place to another, our name tells us that time is neither then nor now; rather, it occurs all at once. To you, our name sounds like the chant of OM sung in reverse.

The Drones: The Holiest of Beings

The first time I saw a drone up close, I marveled at his enormous eyes—so large, they covered his entire head. "Surely," I thought, "there is something more to see when one has eyes that big."

The drones have no stinger, either. They are not made for war. They are made for love.

Often I have lain in the grass with a drone or two on my hand, watching them walk up and down my fingers, in no hurry at all. They seem to be happy wherever they are and within their own contemplative rhythm. Unlike the maiden bees, who scurry to get their tasks completed, drones wander slowly, with tempered curiosity and composure, at one with the world.

Drones make up a small part of the hive's population—about 5 to 15 percent in the foraging seasons. In conventional beekeeping, drones are presumed necessary for one thing only: mating with a virgin queen. Most beekeepers believe drones simply take up space, eating honey that could be better used by the maiden bees or harvested by the beekeeper. Due to their size and longer gestational cycles, drones may also be magnets for parasitic insects, which target the drones while they are still larvae.

Following these beliefs about the drones' lack of usefulness, most conventional and even some natural beekeepers cull and kill the drone eggs, leaving only maidens and a queen. But no matter how many times beekeepers find and kill the drone eggs, the queen continues to lay more to replace the missing ones. Obviously, the hive wants drones.

Why? Because drones carry the genetic line of their hive's queen. As long as drones from this hive mate with virgin queens from another hive, this hive's genetic line continues.

The drones do see more with their enormous eyes. Because bees experience their hive as a unity of shared thought and consciousness, when the drone leaves the hive, the hive's queen and the maidens experience the world through the drone's heightened sensitivity. When a drone flies out to breed or to visit other hives, all the bees of his hive visit as well. Through their drones, the colony perceives the inner chambers of other hives—the sights, sounds, scents, and emotions—and is connected to the wisdom and knowledge of the greater bee community. In this way, a drone's freedom expands his hive's experience and knowledge. Rudolf Steiner, the visionary philosopher-scientist, called drones "the sense organ of the hive" and said they are responsible for communicating to the hive the feeling states in which the hive dwells.

Drones are exquisitely conscious of the sense impressions within their hive. They are also fully connected to the historical context within which all bees dwell, and they transmit this knowledge to the unborn bees through their magnificent song. The drones' song tells new bees about the journey of bees throughout history, from the ancient past to the present, and how the bee kingdom moves forward into its future. Their song describes the spiritual and functional purpose of honeybees in the world.

When the holy drones sing to the pips, they are much like people of Aboriginal and African cultures who sing ceremonial birth songs. These tribal people understand that the birth songs welcome babies into this world and convey important knowledge, telling them where they have come from and how they and their tribe will move into the future. These tribes believe that people who are born without hearing their birth song struggle throughout their lives, because they are untethered and don't comprehend where and how they fit in the world.

While the drones sing their ancestral song, the babies are also surrounded by a second song, sung by the maidens: the hive's song. This vibratory lullaby permeates the eggs in their cells and speaks to them about life within the hive in the present moment. Bees need to know the individual tasks they will take up inside and outside the hive—the role of bees in the world. The song of the maidens tells the bees what they will do once they are born; the song of the drones tells them why they will do that.

The drones are the only bees who sing the Song of Ancestral Knowledge. The drones convey to the hive's future forager maidens the intelligence they will carry out into the fields and bring to the flowers. If the maidens don't hear the drone song and don't acquire this knowledge, they will be less able to fulfill their role in the fields and on earth.

In a robustly healthy hive, each bee resonates with the drone song. Inside the hive, the drones bring forth the balanced sound that vigorous hives make—the sound of healthy, exuberantly alive hives.

There is another clue that hints at the unique role the drones play in hive life: unlike maiden bees, drones are able to visit other hives. Each hive has its own distinct scent, which comes from the queen's pheromone. All bees in that hive carry her scent on them. If a maiden bee from one hive goes wandering to inspect the pantry of the hive down the lane, the second hive's guard bees know immediately by her scent that she is not from their hive, and they'll chase her off to prevent her from robbing them. But drones can visit any hive. They land at the entrance, stroll past the guard bees, and head inside. The guard bees step aside and let them enter, even though they know the drones are from another hive.

Why would the guard bees do that, and where are the drones going? Common sense would say that the first place you wouldn't want strangers in your home is the nursery, but that's exactly where the drones go: into the brood chamber where the developing pips are. They join the other drones in two tasks: providing warmth to the brood and singing the drone song.

The drones maintain the Song of the World, a worldview shared with all hives about the role of bees in the cosmos. They hold and create the prayer that carries the hive along. If a hive were a ship, drones would be the keel. They are at the center of the healthy functioning of the hive.

IN OUR OWN WORDS

Drones are the freedom of the hive.

When the drones fly out, their senses open to the world. As drones encounter the world, the queen and all the bees of their hive experience the world through the drones' heightened awareness. When drones are removed, the hive has no window of perception to sister bee communities.

Because a hive shares consciousness, when a drone visits other hives, the queen and all the maidens visit that hive as well. Through the experience and senses of the drones, the colony perceives the inner chambers of another hive—the sounds, scents, and emotions; the flurry of chase; the color of the light that shines from the lumen. When a drone mates with a virgin queen in the lumen (a holy mating place where light emanates from the earth), the entire hive has a perception of that union. The free visitation of drones to other hives connects each hive to the wisdom of the greater bee community.

Drones sacrifice themselves. They are the holiest of beings.

They make the prayer sound within the hive and are not distracted by tasks. They sing, and their sound fills us with love. Their sound pulls the babies, as they are being born, through the light and imprints the unborn bees with the vibration of creation. The song, the Arc of Creation, moves through the hive like a prayer, becoming part of the vibratory being of each bee.

The creation song tells about a world with the sky and the earth and a horizon between. The song encodes each larvae with "how to bee," overlaid with a blueprint of how the world comes into being each day and how the hive helps carry the world into the future. When the larvae are mature, the creation song calls them out of the cell and stimulates the sensory organs of the baby bee through a linkup—an invitation and an answer, a call and response—as they leave the cell.

Within the creation song is an elaborate and precise communication about the organization of minerals and a map of the relationship of the mineral forces contained in pollen. This chemical language reveals the right relationship between the minerals, with a specific awareness of the hexagonal silica. The sound the babies have been immersed in describes an imagination of a hexa-gon, even as they are being formed inside one. This image, as they grow, will come to have great meaning to them.

While the drones sing, the baby bees are also imbued with a sense impression of the practical tasks of the maidens through the constant hum and vibration of each task within the hive. The movement of the comb and the progress of individual tasks occur-ring upon it are vibrationally conveyed to all bees in the hive. From the maidens, the babies hear and understand the industry of their daily tasks—each activity with its own rhythm and vibration. From the drones, the babies hear the past and future of the bees.

Before and as they are being born, new bees hear a combined harmony of two songs that make up the Creation Song: from the maidens we hear, "Come and join the work," and from the drones we hear, "Come and join the world."

The Queen: The Sun in Our Constellation

Right now in my bee yard I have a hive that just lost its beloved queen. I don't know what caused her loss, but I do hear the effects. When I put my ear to the side of the hive, I hear the colony mourning. Instead of a vibrant steady hum, the hive's song wavers with questioning trills and piercing, high-pitched stabs amid an undercurrent of moans: "The queen is dead. Woe to us all. The queen is dead."

A nearby hive that recently swarmed is releasing newly birthed queens left and right. I've found two of them so far, unmated as yet, looking a little dazed as they wait out the few days before their mating flights. These unmated queens are fed and watered by the maidens but are not recognized as special; they have no task yet. The unmated queen is just another bee. Once mated and accepted by the hive as the new mother, she becomes a queen, and the hive is referred to as *queenright*.

The queen is the most significant bee in the colony because she is the hive's reproductive force, the mother of every bee born into that hive until the end of her reign. She continually fills cells with eggs, ensuring the ongoing life of the colony. In springtime, a strong queen builds the population quickly and keeps it stable through each season. Queenright hives are joyous, steady, and filled with purposeful activities.

The queen lives inside the hive and never sees the light of day, except on her mating flight and during swarming. Swarming is how bees create new hives, reproducing in the larger sense. Bees do not expand the number of individual bees through swarming (which is an in-hive process); instead, they increase the number of hives in a geographical area. As an indicator of successful winter survival and good spring health, a portion of the colony will prepare to leave the old hive behind and build a new hive, thus adding a new colony to the local area. In anticipation of their coming adventure, healthy colonies fill their hives with pollen, nectar, and thousands of bee eggs as a gift to the bees who stay behind and support the old hive.

When the colony is ready and the weather dry and sunny, about two-thirds of the colony departs with the queen. Within a few days of leaving, the swarm will establish itself in a new location and set about constructing another hive. Once the maidens have built enough new comb, the queen begins filling the cells with thousands of eggs to quickly bring her new hive up to a healthy population level. Each spring she'll repeat this process—swarming and relocating, leaving behind the old hive populated by the next generation.

What happens to the original hive after most of the bees and the queen have flown off to make a new home? The bees left behind will continue with their tasks, but with only one-third as many bees. It takes a few weeks to ramp back up to full capacity and production. Each day, more pips will hatch, and the population will grow.

Before leaving, the swarming bees made queen cells—special places for the baby queens to gestate. These cells are structurally different from the narrow, horizontal cells within the comb that hatch out maidens and drones in twenty-one to twenty-five days. Queen cells are long, peanut-like appendages that hang vertically off the edge of the comb. One of the eggs within them will become the hive's new monarch.

The developing baby queens are fed an exclusive diet of royal jelly that hormonally changes their form and purpose to that of a sexually functional queen. Newly hatched queen bees are already much larger than the maidens and drones, and their hormone-rich diet brings them to size and maturity sooner than other bees—in just fifteen days. Young virgin queens spend a few days wandering about the hive as their bodies mature enough to mate and as they await the perfect sunny, warm day for their first nuptial flight.

On an unmated queen's nuptial flight (sometimes she takes more than one), she mates with twelve to twenty different drones from other hives—never with drones from her own hive. Once mated, a new queen returns to the old hive, where she searches out and slays all the other unhatched virgin or other mated queens who, like her, are just returning from their marital flights. Typically, only one queen will survive. (More on that in chapter V, which talks about swarm behavior.) Once all but she are dead, she ascends to her role and becomes the queen. Instantly, the colony organizes itself around her.

Within weeks this colony of bees left behind during the last swarming will become a new entity in itself, because the offspring of this new queen carry partially different genetics and ancestry into the hive. The drones from other hives she mated with on her nuptial flight(s) provide the colony with a wide diversity of bee traits that contribute to that hive's biological continuity. This mixture of lineages allows for different traits to come forward as needed. For example, if the weather becomes colder than normal, one lineage may have special knowledge of how to keep the cluster and larvae warm, thus saving the colony from freezing. In another situation, a long drought may call upon the lineage of bees who are adept at finding and bringing water to the hive. A third lineage may be the best pollen gatherers ever. Depending on the genetics of the new queen and the drones she has mated with, the personality and behaviors of the hive may change. In this way the hive itself continues living, even though the members of the colony change from year to year.

The new queen immediately gets to work restoring the hive to full capacity. She lays up to two thousand eggs each day (except in winter) and can do this for five to seven years. Within the hive, the queen's cluster of handmaidens surrounds her in a circle of love and appreciation. They care for her in every way. They groom her, prepare cells for her to lay eggs in, and feed her the beneficial royal jelly she needs to maintain her fertility and health. As the queen moves from one cell to the next, lowering herself into each cell to deposit yet another egg, the handmaidens care for her. She is always in the center of the hand-maiden circle. They radiate out from her like the petals of a sunflower. The queen has a distinctive vibratory sound upon the comb, because she is always in the company of her handmaidens.

Each queen also exudes a unique scent that identifies her. Her scent has multiple functions: It suppresses the reproductive ability of any other female bee. It identifies all the bees who live in that particular hive, helping guard bees decide who belongs and who doesn't. It discourages drones from the same hive from mating with a queen who is also their sister. It encourages comb building and other productive activities. And perhaps most important of all, her scent establishes cohesiveness and calm throughout the hive, bringing contentment to all.

The queen's scent rises from glands in her head that exude a pher-omone called queen substance. The queen's handmaidens groom her by licking her all over, and in doing so, they pick up some of her pheromone. Bees feed each other by exchanging food between their mouths, and in that action they also transfer the queen's scent from one bee to the next. By this process, each bee takes on the scent of his or her queen. The entire colony becomes enthralled with her scent. This effusive queen aroma constantly wends itself through-out the hive, signaling her ongoing prolific fertility. Toward the end of her years, the queen's fertility drops off, reducing the strength of her pheromone and scent. When the maidens grasp that the queen's fertility is declining, they will construct a replacement queen cell in the center of the comb and care for the new queen pip until she is born.

After fifteen days, this new queen will hatch, mate, and begin laying eggs. Most beekeepers say that only one queen can exist in a hive and that the hive will kill off the old queen. However, I and other natural beekeepers have seen colonies where the old queen was allowed to stay in her hive for the duration of her days. When the new queen took over the role of egg laying, the dowager queen moved apart from the cluster and continued on, fed and warmed, until her natural death.

Once I opened a hive and was startled to see an old queen. I knew she was older because she was no longer fuzzy bottomed, having worn off her abdominal hairs slipping in and out of thousands of comb cells. I also noticed that none of the nearby bees were in the protective circle around her. She looked up at me, both of us surprised; she then dipped down between the combs. I immediately put the top back on the hive while I considered what to do. A few minutes later I opened the top of the hive again, and the dowager queen scooted up from between the combs, perched on the very top of the hive for a moment, and then set her wings in a whir and flew off toward a field, where her demise would occur in the warmth of the sun.

The Effect of Commercially Bred Queens

These days, 90 percent of the queens in conventional beekeeping are raised by a small handful of breeding companies, who breed for desirable traits like cleanliness, low propolis production, less desire to swarm, docility, or lavish honey yield. Sadly, these selection criteria may also unknowingly breed out traits needed for natural disease resistance, robust survival, or as-yet-uncalled-upon traits that may be needed in future situations, like changes in local weather.

Most domestic bees are raised outside the bee buyer's local area and then shipped all over the country and even to other continents. A beekeeper who lives in the rainy Pacific Northwest region, for example, may like the idea of Texas bees, who are big honey producers, and place an order. But those sun-loving bees may struggle when they have to live through seven months of rain in Washington. Some small-scale breeders are starting to raise survivor bees local to their region and letting them mate with feral drones. If you have that in your area, they are worth seeking out.

Purchasing queens narrows the gene pool, as queens produced by artificial insemination are usually mated on an assembly line with only one drone or only drones from the same lineage. Those drones, you can be sure, were not selected because they were the fastest and strongest, which is nature's criteria. The lack of genetic diversity brought about by commercial breeding diminishes the internal workings of the hive. Feral breeding, with one queen mating with a dozen or more wild drones, allows a hive's residents to experience an abundance of

traits and behaviors. We humans don't understand the variations necessary for all situations, and we do the bee family harm when we limit
their diversity.

Replacing a queen, which conventional beekeepers commonly
do, is traumatic for a hive. The queen is the conduit for the hive's
life force; it cannot live without her. She is the central focus of
the bees' dedication to their family, like a beating heart. Their
connection to her is crucial to their survival and well-being. Yet
conventional beekeepers replace their queens with a fresh new
one every year, not understanding the significance of her loss to
the colony.

In a hive's natural annual swarming, the queen's fertility is rekin-
dled for another year (see chapter V). Because many conventional
beekeepers prevent hives from swarming, and thus keep the queens
from maintaining their fertility, they need to replace the queens each
year with freshly mated, newly purchased ones. In a terrible process
called "pinching the queen," the old queen is found, plucked out
of her home, and crushed to death. The new queen, in a cage to
prevent the colony from killing her, is unceremoniously placed into
the hive. The new queen is not related in any way to the colony; her
scent is completely foreign to them. If she were not protected by a
screened box when dropped into the hive, she would instantly be
labeled a stranger and killed. It takes days for the bees to acquaint
themselves with the new, off-smelling, boxed-up queen before they
accept her.

In his book *Wisdom of the Bees: Principles for Biodynamic Beekeeping*,
Erik Berrevoets writes:

> Conventional beekeepers see the introduction of a queen
> bee into a colony as somewhat like introducing a dairy
> cow into an established herd. [Rudolf] Steiner's research
> showed, however, that the relationships among bees in
> a bee colony are much more intricate, and that a queen
> bee can be compared more appropriately to a human or
> animal organ. In fact, for the bee colony the effects of
> introducing an artificially bred queen bee is more like an
> organ transplant.

IN OUR OWN WORDS

Inside the hive, the queen wends her beautiful fragrance, bringing us all into a harmony made of scent and sound.

Her scent is an elixir that contains all we can know of her. Our matriarchal lineage disseminates itself inside us. The merging of the lives of our fathers sings to us. The adventure of them finding each other plays inside us, a story we love to hear. In her scent we feel the sun call her to climb the sky, to reveal herself to the plants, and we hear how each of our fathers gave chase and sealed our presence with their kisses.

Each queen's scent differs from the next by her story. Through her we are included; we are the resulting joy. We know her love for us by her telling of how we came to be, how our fathers and mother, and our fathers' fathers and mothers, and our mother's fathers and mothers, all expended the greatest effort to bring us into the world.

This scent surrounds us and speaks to us every day, telling us how loved we are, and in return, we want nothing but to honor them by being in service to the hive. Her scent calls to us saying, "You come from love. You live in love. You are the love. Blessed are we all who sing this song with our breath."

When beekeepers replace a queen, thought is not given to how this affects us. The loss of a colony's queen is devastating to the unity of the hive. We grieve inconsolably in her absence, and then a stranger is thrown into our midst. The unknown queen from faraway arrives while we are in mourning. We are no welcoming colony. She does not know us, and we have no knowledge of her marriage. We have no embrace for her.

A stranger, she arrives without welcome and is initially scorned and rejected. It is only through our recognition of the complete loss of our familiar queen that we allow this new queen to step into the central role, to lay down the order that is the central architecture of the hive itself. This new queen, who smells of a foreign land, does not know our ancestors, the elements, or the natural history of our sisters and brothers. Our true queen is gone, and without her, our family line is empty.

Pinching our queen is a desecration. When a colony replaces a queen, we do it without violence. Thus, the crown of creation moves through time with assent and purpose and always with love.

The Queen's Walkabout

In the wild, bees move freely within the hive, attending to the comb and each cell. The youngest bees work in the nursery, the guard bees guard the entrance, and the wax makers pitch in wherever new comb is being made. Generally these bees stay in the areas of their tasks, but they all—including the queen—have access to every corner of the hive.

The queen spends nearly all her time in the nursery, or brood chamber, where her contingent of handmaidens cares for her every need so she can focus on laying eggs and keeping the hive's population appropriate to the colony's needs. There are, however, special reasons that can induce her to venture into other parts of the hive, such as having more places to lay eggs. Expanding into a bigger nursery means the birth of more bees and is a good thing.

The queen might lay eggs outside the nursery if the wax comb in the brood chamber is too old. Exposure to chemicals happens frequently these days. Even treatment-free bees may inadvertently bring home nectars, pollens, and floral essences tainted by tiny but significant amounts of chemicals, or the bees may carry chemicals into the hive on their bodies. These chemicals become embedded in the wax. Bee larvae mature surrounded by the comb, and if the cell's wax has chemicals in it, the babies are exposed in utero. The queen intuitively prefers to lay her eggs in the cleanest new combs, so she may wander a bit, seeking cells that will keep her babies safer.

Conventional beekeepers prevent the queen from laying eggs in the honey chamber, because to them, brood mixed with honey is problematic. For this reason, a metal barrier screen (called a queen excluder) is placed between the hive's upper hive boxes (called supers), separating them from the rest of the boxes below. The openings between the bars of the screen are wide enough for the maidens to slip through with their donations of nectar and honey to store in the comb, but are too narrow for the wide-bodied queen to squeeze through.

I don't use a queen excluder in my hives because I believe it's important for the queen to have access to the entire hive, as she does in a feral colony. While it is a bit inconvenient to find bee larvae in the honey area, I find that happens rarely. I also know I don't need to harvest every speck of honey the bees make. I'm frugal in taking too much honey anyway, and seeing bee larvae in the honeycomb reminds me not to take too much. If I find any combs of mixed honey and brood, I leave them for the bees and take honey only from combs filled with just honey.

The only reason to confine a queen to the nursery is to make honey gathering faster and easier for humans. When human convenience is the reason for a certain activity, that activity is usually harmful for the bees.

Another important reason for the queen to travel beyond the nursery is to spread her unique queen scent throughout the hive—an act that supports the harmony of the entire colony.

IN OUR OWN WORDS

All is known by scent. Scent is a descriptive listing of all contained within this body, whether plant or animal.

Within the hive, the scent informs the sound a hive makes as we declare our current state. Scent is an indicator of waning or waxing health, of seasons past and present, of royal fecundity, of gestational flowing.

The queen lays her royal scent throughout the hive. Her scent confirms our ongoing populousness. All are cheered by this.

She is mostly in the nursery, but she sometimes walks through the hive. When the queen visits a hinterland, her scent proclaims her presence and leaves the mark of her royal visitation. It communicates to us the strength of the hive in saying that we could, if we desired, expand the core of the hive—the nursery—beyond.

It is a dreaming and may never be acted upon, but as a dreaming, it asserts the willingness to flourish. This queen's scent throughout the hive says we could, if we desired, become larger. Every bee is joyous in this imagining, as it confirms the hive's ability to expand and grow. This joy is communicated throughout the hive by singing another verse of the Song of Increase.

The Queen's Gift and the Drone's Promise

A drone is born of a queen. A queen is born of the two.

THE BEES

When I first read that a drone is born from an unfertilized egg, I thought I had misread. How can anything be born from an unfertilized egg? When the queen dips her hind end into a cell to lay an egg, she makes a choice. She decides whether to fertilize it with her vast interior stores of drone sperm or to lay an egg that has nothing but her own solo genetic contribution.

Essentially, the drone is an exact clone of the queen, except that he is male and his body differs from hers in form and function. So he is a clone in genetics only. How is that possible? The drone has no father, but he shares grandfathers with some of the maidens.

During the queen's mating flight, many drones donated their sperm, which she then holds within her for her entire life. Each maiden is born of the union of one sperm with one egg. All maidens are half-sisters with a common mother—the queen—and some of the maidens are full sisters who also share a common father—one of the drones.

The female bees, both maidens and queens, have a similar lineage, both born of a fertilized egg with mother queen (X) and father drone (O). The pattern is shown in figure II.1.

But a drone comes of just the one, the queen. He has no father, only a mother. His pattern is shown in figure II.2. When you carry this out further, it looks like the image in figure II.3. Once you see the numbers of bees in each generation, you may recognize a pattern. When we add each number to the previous number, each subsequent number is the sum of

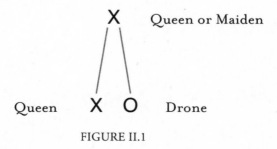

FIGURE II.1

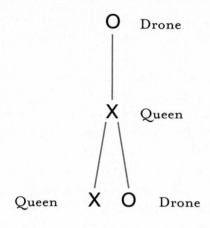

FIGURE II.2

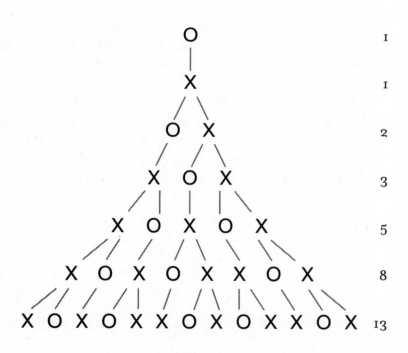

FIGURE II.3 (NUMBER OF BEES IN EACH GENERATION)

the previous two numbers: $1 + 1 = 2$, $2 + 1 = 3$, $3 + 2 = 5$, $5 + 3 = 8$, $8 + 5$ = 13, and on. This is the Fibonacci sequence, a mathematical code that is found throughout nature, like in the spiral of the sunflower, the sequential branching of ascending tree limbs, and the scale pattern of pine cones. It is the basis of the mathematical formula called the Golden Mean or the Divine Proportion. This numerical sequence expresses itself in the geometry of crystals, the unfurling of a fern leaf, and the chambers of a nautilus shell. It is evident in the shape of the cosmos and has been used to create mysterious architecture, such as the Great Pyramid at Giza and the Great Mosque of Kairouan, as well as extraordinary classical music. The mathematical sequence even shows up in our perception of beauty.

When I first realized I was seeing the Fibonacci sequence in the drone's family tree, the image stayed with me for days. I knew it was significant, but I didn't immediately understand why this magic number was present in the drone lineage. Queens and maidens are, after all, only one generation away from the same sequence.

A few days later the bees explained how the drone is the key pivot point that ensures the preservation of the queen's line. The drone does stand alone—apart from the rest of the females and singular in relationship to the queen who bore him. He is the one who carries his queen mother's ancestry and genetic material forward, even in her absence.

The Fibonacci sequence is only activated when a queen—or in rare situations, a maiden—produces a drone. The entire aspect of sacred geometry comes into play only when a drone is born.

As described earlier, normally when the queen grows old and slows down her egg laying, which can happen in a feral queen after five to seven years, the colony readies a cell in which the queen lays the egg that will grow into her replacement. Usually this happens in time for the maidens to raise a new queen before the demise of the old queen. This new queen egg hatches, the dowager retires, and the new queen continues the lineage of the prior queen, ensuring that the colony and its genetics survive. All is well when this happens.

But sometimes the colony loses the queen unexpectedly, before the hive has prepared for her replacement. The hive or the brood area may have inadvertently been damaged or chilled, and no eggs are suitable for becoming new queens. This hive, alas, will perish, and all the queen's line will die with that colony.

As the colony mourns the loss of their queen, a maiden bee may step forward and call upon her virginal reproductive capacity to take the place of the missing queen. Somewhere in the Unity's consciousness, this maiden recognizes the colony's situation and is compelled of her own volition to alter her structure and turn herself into a temporary queen. She becomes fully capable of laying eggs, but only those of drones. Infertile, a virgin, she cannot replace the queen and provide the colony a future, but she can provide a way for the colony's knowledge to continue even after its demise.

Parthenogenesis is the word that describes the ability of a female to lay a viable egg without that egg having been fertilized by a male. It is a rarity in the natural world. The word comes from two Greek words: *parthenos,* meaning "virgin," and *genesis,* meaning "birth." The maiden, when called upon, enacts this virgin birth and sends these last representatives of her line to carry their colony's knowledge into the world.

IN OUR OWN WORDS

Part One: The Drone

Every hive is a compendium of knowledge. Every hive has experienced slightly different situations and learned to discern and intuit various conditions in ways that have allowed the hive to survive and progress. For that reason, each hive is singularly important.

Even when a colony loses its queen, her line—the living bees she leaves behind—still bears knowledge unique to her colony. In the case of a queenless and dying hive, great effort will go into raising more drones to carry on her line and to guarantee that this hive's knowledge continues. A host of drones is made ready, and when these drones mature, they will leave the hive to spread their seed. And so the line is preserved.

When the queen is lost, the colony's knowledge continues by virtue of a maiden who is called, in the queen's absence, to lay the unfertilized drone eggs. When the eggs hatch, the drones go out and mate with new queens of another tribe. In this way, the knowledge contained in each line continues and always moves forward, even in the loss of the queen mother and the subsequent loss of the entire colony. The drone eggs are fundamentally present in every maiden to ensure the line's survival.

When a queen lays a maiden egg, she says, "These are of my union." These eggs are for the hive—for this hive and all the good this hive does.

When the queen lays a drone egg, she says, "This is my gift." The drones that hatch from those eggs go out and help populate other hives. When a queen lays a drone egg, it is the queen's gift to the whole bee kingdom. When a drone flies out and finds a virgin queen and mates with her, his queen's gift is received and her contribution to the eternity of bees fulfilled. The old queen will never see her grandchildren, yet her colony's knowledge carries on, seeding the kingdom's future.

Part Two: The Maiden and the Out-of-Season Drone

The code we speak of is God's design for multiplication. Each number has a precise foundation built upon itself that supports it before it progresses to the next. Multiplication explodes into being. "Here is one. Here are many."

The code, however, is the natural way of bringing numerical growth in such a way that life emerges out of itself. This pattern is far more frequent than imagined. These numbers have a security to them; they are just and tempered.

The code comes forth in science and mathematics, saying, "God's hand is here." Protect with all effort where the code expresses. These expressions are crucial components of evolution. Such knowledge is needed to counter subtraction in a world gone awry.

Our last gasp of life that comes of the drones is essential because it carries forth hidden knowledge. When a colony is about to die, each drone becomes the repository of all that colony has known, and the knowledge is woven back into the field. When a colony seems to be failing, let the colony focus itself upon these drones, for it is by this focus that their secrets are shared so they don't remain hidden.

Often these drones are presumed errors, but they are not mistakes. They are part of our plan. The bounty of a failing hive is not the left-behind stores; it is the repository of knowledge held in the drone who is freed to mate himself into the colony that needs what he knows. It is out of season, in late summer and fall, when these drones are most needed. During normal spring mating times, there are plenty of drones to seed.

Sometimes we feel the need for more desirable traits that further our colony's ability to respond to challenges. If a nearby colony has traits we need to express, a proper combining of specific knowledge ensues. We invite the old queen to "abscond," to leave her home, while bringing into being a new queen. The old queen's departure is not a mistake; it is simply a shift in course for the benefit of the Unity.

Late-season drones are meant to find a late-season queen and marry. If the drone's colony is failing, it has been deemed of more use as a father to a next generation. Consequently, all focus of the maidens goes into birthing these traveling drones.

The code continues. It emerges through a maiden no longer barren, who begets in the drone the sum of the hive. He then carries his exquisite and precise presence into a waiting world.

A maiden who carries the seed of the drone is thought to be unmated. She has been awakened—not by mating, but by a sovereign language within the code that calls her forth. Pregnant with a virgin birth, this expression of God's love speaks through her, and thus, she bears the drone into being.

The Nursery: The Heart of the Hive

When I open a hive, the bees continue working with barely any acknowledgment of my presence. There is, however, one area of the hive I nearly never enter: the nursery, where the pips develop. I purposely don't spend much time in the bee nursery because it's the most private area of the hive. The brood chamber is the hive's uterus. I don't believe we should open the brood chamber, because it causes concern among the nurse bees. Seeing the bees concerned always gives me pause and makes me question if I need to be where I'm not wanted.

In the brood chamber, I've seen nurse bees stand over the eggs like a shield, protecting them with their bodies. Only with persistent urging will they reluctantly move off brood cells. An opened hive doesn't retain enough heat for the pips, so the maidens collectively position themselves atop each egg, warming the pips with their body heat to prevent them from getting cold. What good nursemaids they are.

Joseph and I were once called to rescue a colony whose willow tree home had split apart in a storm. Half the tree lay shattered on the ground; the other half of the split trunk was leaning in a standing arch supported by one thick limb. Underneath this curve we found many pieces of comb scattered on the ground, and some still adhered to the bottom of the limb. We wanted to save as much comb as possible to reconstruct (as well as humans can) a new home for the colony. I picked up many large sections of honey and pollen, but most important were the egg-filled brood combs that held the next generation's bees. If we could save the babies and find the queen, the colony might have a chance.

All the brood comb was thickly covered with bees. When the tree had fallen apart and scattered the comb, the bees had immediately attached themselves to the brood comb. Chilled brood is dead brood, and the bees were trying their best to keep the unhatched larvae warm and protected from the cold air. The honeycomb and pollen comb pieces had nary a bee on them. Every available bee was on blanket duty, insulating the brood.

Every time I thought this bee rescue was futile, I'd find another chunk of brood comb covered with a blanket of bees. They'd been tossed about and rained upon, yet they continued trying to save the brood. I saw the commitment these little bees had to protecting their young, even in the face of overwhelmingly difficult odds.

We set Joseph up on a makeshift table with his homemade emergency frames, a sharp knife to cut the comb into the frame's shape, and string. My task was to hold the comb steady and use a feather to move the bees off the comb long enough for Joseph to tie it into the frame. That was hard to do. Every stroke I took parted the bees, but they immediately refilled the empty section, continually trying to keep the babies warm. They had a mission to preserve the viability of the baby bees and were diligent in tending to that task.

All afternoon I worked under the arch of the enormous, still-standing half-trunk, picking comb off the ground and removing sections in the split wood over my head. After five hours I told Joseph I thought we'd gotten it all, and I stepped out from under the limb. I had taken only a few steps when suddenly the huge limb supporting the trunk broke, and the trunk fell flat onto the ground, exactly where I'd been standing.

The timing was so precise, maybe fifteen seconds after I walked away. I stared at the spot I'd been in all afternoon, now covered by a ton of solid tree trunk, and felt great relief that the guardian angel who held up the broken limb for all those hours had supported it until all the bees and I were safe.

Joseph and I worked all afternoon in the blowy rain. We did get the hive moved, and thankfully, this lovely colony, which I guessed was a few decades old, survived.

IN OUR OWN WORDS

The brood is the heart of the hive. We pour our love into the new ones, surround them with love, and beyond the nursery, surround them with industry. We work and we love; they are one and the same.

At the right temperature, certain chemical and alchemical processes take place. If the heat is too low, the pips are compromised. If the temperature is even a little off, they may survive but will not be as strong and not as responsive to challenges.

This is very important: the nursery should not be disturbed. We have built the nursery structure in place, and we work to keep it at an exact temperature. When the hive has been opened, we have to repair it to enclose the womb again.

When the temperature is right in a developing bee's cell, a melding takes place in the bee's brain and nervous system. A bee born from these perfect conditions is strong, well protected, and quick—a responsive and receptive member of the community with perceptual abilities that allow the bee to sense and communicate conditions throughout the hive. These bees perceive changes and immediately let the hive know and take action to remedy what needs attention.

A bee not melded is slower. Still functional, this lesser bee can sense and act, but with less perception and initiative. Fully melded bees are part of the collective and can initiate appropriate responses. Less melded bees can only act as followers, and thus, the hive suffers.

You may look at the comb and the honey, but when you near the nursery, please let it be. This is our most delicate area, and it's easy to cause damage without knowing you've done so.

The Pips: Before We Have Thought, We Have Knowing

One sunny morning in early July, I was about to harvest some honey from my bees. This colony looked large and robust, and I was pretty sure their top box was full of honey.

The colony lived in a Warre hive, a hive type that mimics a hollow tree. In my apiary the Warre colonies are rarely disturbed, so the bees build as they choose, which sometimes means curvy comb under the simple top bars. I carefully pulled a thin piece of wire between the uppermost box and the box beneath it to break the propolis seal the bees had made to hold the boxes together. Then I lifted off the top box and placed it on a tray next to me.

I expected the top box to be all honey, and lower, in the second and third boxes, would be the egg-filled brood chambers. That's how they normally are. However, I saw that a small part of the brood chamber in the second box had extended a few rows up into the top box. Oh dear. This was unfortunate. I'd already cut the boxes apart, and in doing so, I'd nicked the sealed wax cover off a few pupating cells, exposing the white, unhatched pips to the harsh light of day.

I felt bad. The poor little bee grubs had been close to hatch time, but now they were damaged and didn't have a chance. The pips swirled in their cells, surprised and unprepared for this too-soon birth. There was really nothing I could do because the injured pip would be rejected and would die, even if I tried to put them back together. Next time I would be more careful.

Pip is the word the bees use to describe unhatched bees of any stage. *Seed* is an egg before it has been fertilized and placed in a cell; after fertilization it becomes a pip. Once the eggs are hatched, they are called maidens or drones.

While I devote much time to bees, I don't often see the pips because they spend all their time in the cells, developing. The queen lays a tiny egg in the bottom of a cell, and three days later the egg hatches and becomes a wormlike larva. After six days, the pip spins a cocoon, and over the next weeks, all the parts metamorphose into a beautiful bee hatchling.

The queen determines the sex of the seed. Laying an egg that is not fertilized (a drone) washes the channel inside her and keeps her reproductive system healthy. Laying a drone egg allows her to take a brief rest from the constant rhythm of merging sperm and eggs. Drone laying is a time to breathe easy, to use less effort and action.

Much magic happens between the laying of the egg and the pip's birth. The bee birth story is magnificent, involving every member of the colony.

IN OUR OWN WORDS

When the queen lays an egg, she brings together the seed and the sperm. She holds within herself hundreds of thousands of seeds, and she has already collected enough sperm to fertilize them, though this is not done all at once. Rather, it is done daily in warm seasons, with a thousand or more seeds fertilized each day for many years. She brings together sperm and seed just before she lays the impregnated seed into the cell.

A constant pulsing wave moves through the chamber of the queen's abdomen as the seeds are drawn down and through and then kissed by the sperm. Thousands of radiant threads of light emit from the surface of each egg, all the way around. These glimmering filaments of light hold an electrical charge that's incredibly attractive to the sperm, who seek the egg the moment it's released.

The strands call the sperm and wait expectantly for it to answer. The light the seed emits through its filaments is a spectrum common only to this purpose. The filaments channel the sperm so that they move forward, head first, toward the seed—the only position from which they are able to penetrate it. At last, the head of the sperm surges straight into the seed, and in that moment, the egg becomes fertile.

The sperm enters the seed bearing a blessing, a waking instant knowledge that immediately instills a sense of structured order and direction into the stirring complexity within. In the moment of union, the egg awakens to the knowledge of how it will grow and become part of the community of bees. At its time, the newly joined seed enters the chamber like a wave meeting the shore. Gently held, caressed by the wave, the seed flows to the opening portal, where the queen delicately places the blessed new egg into the cell.

The making of the egg is a roiling process. The universe focuses so that the sum of all energies is present in the pip. Nurse bees move through the corridors, first feeding royal jelly to the babies in their cradles, and later, as the babies mature, they feed pollen.

Within the chrysalis we eat the encapsulated light of pollen. The light-filled pollen informs us, shapes us to the world outside the hive. Before we have thought, we have knowing.

The pollen has been specially prepared for pips. When we eat the pollen, we experience light in many forms, as you would taste blended flavors. The light signature of each pollen expresses itself as an individual taste, a more tubular sensation, separate from each other, yet fused, like a bundle of fiber optics.

The specially prepared pollen speaks of the present. The pollen explains this place, the terrain, the seasons, this moment. It tells of the scents on the winds and even the miasma of harm that may be present in the area.

Royal jelly enables the infant bees to decode the Song of Creation, which is the song of past and future. A drop of royal jelly is fed to the pips in the first few days. Through royal jelly, all hives are interconnected, and we see the consciousness of the Overlighting Being present in each hive. Royal jelly is *inter*hive, while our pollen is *intra*hive.

The maiden pips in the cells move in a repetitive coiling, like the dance of the dervishes, turning in all directions, but mostly in the direction of the earth's rotation. In this way, we familiarize ourselves with the roll of the planet.

Inside the egg, our pips are putting together their first thought with a language given to them by the royal jelly, the pollen, and the songs of the drones and maidens. Bee language is a scent on wind, harmonious existence, the yeasty taste of the field pollens. The pollens are mixed because that expands our palate and encourages exploration in mature bees. Pollen tells us, when we go out to harvest, "Find what tastes like this."

The eyes of pips are not yet developed enough to see light, yet their brains have a perception, an imagination, of light. They have a framework of knowledge of light even before they go out into the light. When they come into the light for the first time, all their colors come together.

When we feed the prepared pollen to the babies, we feed them light. The little being that is born comes through the Arc of Creation—a wide opening filled with light—and the inside of the hive glows with light. Coming through the arc is something each bee must do on her own. A blessing is given to each bee as she is born:

> We come through the door with wings folded
> tight to our bodies, white as day.
> The first breath colors us. We enter the world
> as the world enters us. Life and light of Creation.

The pips know it is time to be born because they feel a tremor go through their muscles, and a linkup happens. This tremor confirms each part is connected within them. As the mounting wave initiates forward movement, the pips chew away the caps that hold them in.

A loud crack, and then the whole inner sound of the hive. As the cell seals break open, the pips sense vibration, and they immerse themselves in the sound. Inside the cell, with everything pressed tight to them, they had no distinction between themselves and the sound of the hive. Passing through the cap of the cell—shaking, drying themselves, and cleaning themselves—their perception of the sound is completely different from when they were inside the cell.

In the cell, the pip and the cell and the hive are one. All the energy of the hive goes into nurturing and creating this single being who has no sense of singularity. In the cell, she feels the sound, and once she breaks through, she helps make the sound. She goes from being the focus of the hive's creative energy to becoming part of that creative supportive force. Inside the cell, she is a receptor, drawing energy to herself. When she comes through that door, she begins to sing.

The right to life is earned. Birth is difficult, the entry intentionally challenging. Birth is nature's gauntlet through which one passes alone and, upon reaching the portal, is given breath. The challenge sorts weak from strong, damaged from vital—measuring the life force.

In the hive, we cannot help the birth. We each must bear the measure. Weakness is continually culled from the hive. We note the crinkled wing, the hobbled knee, the effects of poison, and we remove the malformed pip.

Poisons have no place in this world, where the delicacy of connections so fragilely frames the underpinnings of our evolution. In our world, that which drains life would not persist. There is no good in that which causes life to perish. Such practices are abhorrent and bring despair. Life is the celebration of the joyful gift. That which stifles life or riles its entry does not belong in the flow of colonies.

It is not that the deformed body is wrong; it is what caused the deformation that does not belong in our midst. An environment filled with poisons that block or alter DNA's messages prevents our spirit-filled bodies from transmuting our beings on the altar of evolution. We move to the light, and it infuses us with a progression that opens us to our highest joy.

III

The Song of Communion

How Bees Create a Perfect Home

Before I became involved with bees, I had no idea how complex their lives were. The sum total of my bee knowledge back then was that a bee gathers honey in the daytime, sleeps in the hive at night, stings when upset.

When I became a beekeeper, the opportunity to spend time with my bees delighted me, and I spent many, many hours in their company. Through this observation and, finally, through their generosity in opening their lives to me, I began to see the immense complexity of their culture and society.

In this chapter you'll read about how the bees put together their hive, their exquisitely ordered home, which functions as a living body of the Bien. You'll learn the significance of scent, sound, comb construction, and touch; the function of propolis; and how the bees protect and defend the hive. Disruption in any one of these areas could cause the colony to fail. When everything is done right, the colony expands and thrives. The air around an exuberantly joyful hive shimmers with the blissfulness that emanates from the hive.

IN OUR OWN WORDS

What makes bees happy?

The shape of scents, the sweet of nectars, the sun's prismatic light, our cumulative joy.

What makes bees strong?

First is the vigor of the queen's emanation. All bees are within her orbit. The hive contains her beloved fragrance, which refreshes and enlivens us.

Next is the integrity of the propolis seal. Our medicines are broadcast on the air that surrounds us, and our daily health is dependent upon our connection to our hive air.

Important, too, is the light that emanates from the pollen we've brought in—the bountiful wealth for our young. Add the sanctity of the brood area, that all may flourish within it. Then add the quality of the nectar, that it may provide us with superior nutrition to keep us healthy.

We are most joyful when our hives are high in the trees because we are creatures of the air. When we fly out of our high home, we immediately see the landscape, the light all around. We notice what calls us by the scents on the wind, the breeze carrying news of what is in bloom. High in the air we are in our element.

Comb: A Language of Thought and Feeling

When I open the hive, the honeyed scent of warm beeswax wafts into the air around me. Comb scent always carries me into bee reverie. The scent wafting up and out of a hive is intoxicating—a mixture of nectar, resins, and the glandular aroma of the bees.

The few bees who come to the top look at me with trust and curiosity; we mean each other no harm. I move slowly with the delighted smile of a five-year-old, so pleased am I to see them.

Each time I lift a bar of comb out of the hive, I marvel at the architecture the bees' Unity consciousness calls them to create. Those of us privileged to gaze within the inner workings of a successful hive see a world of sensuous, amber-colored beauty. The combs in naturally kept hives undulate in exquisite organic patterns.

Certain styles of synthetic hives give bees fewer options for how they build their comb. I don't use hives with rectangular frames that stipulate comb shapes that run counter to the bees' intuitive desire. In the wild, bees naturally build their combs below and attached to a wooden surface, such as a hollow tree. The hives I provide offer a similar space, a simple row of parallel wooden bars perched at the top of each hive box, with open space beneath. The bees build combs that hang in graceful rounded shapes beneath the bars.

Langstroth hives have rectangular frames with reusable plastic foundations that are pre-pressed into cell shapes. The theory is that because the cells on the foundation are already made, bees won't have to waste time making wax and building comb; they can get on with the more profitable work of making honey. However, I believe if bees are designed to make wax (as they are), we ought to let them do it.

Unlike any other animal, honeybees create the complete inner structure of their home with a substance they manufacture from their own bodies. The wax secretes from glands in the lower abdomen and hardens into tiny plates that are ready to form into the comb. Young maidens, twelve to eighteen days old, produce the wax, which is then harvested by comb-building maidens who carry the wax chips with their mouths, adding an enzyme that softens the wax so it can be chewed into the proper shape. If more wax is needed than can be provided by young maidens, older maidens can reawaken this ability and contribute wax, too. What a remarkable ability they have, to turn back time when needed.

Comb is constructed in long parallel sheets wide enough for two bees to pass each other back-to-back as they move across their respective panels. The space in between the comb panels is known as *bee space*—a precise measurement that is fairly uniform throughout the hive.

Newly made comb is pure white and light as air. Once they begin to use the comb, bees add more layers of wax and even rim the outer edges with propolis to protect it from damage. Over time the color darkens, and the cell walls become thicker. Unfortunately, the wax is also highly absorbent. When bees are exposed to chemicals during their foraging or, even more obviously, when chemical treatments are used in the hive itself, the wax absorbs these chemicals, which then become ongoing pollutants. The pollutants continue to expose the bees to impurities and toxins. Moving on polluted combs is, for bees, what sleeping on dirty sheets every night is for humans. Even when the beekeeper is a chemical-free, treatment-free beekeeper, it's hard to keep toxic chemicals out of the hive. For this reason, I often rotate out the older comb, so the bees will create new clean comb for their home.

Besides providing an internal structure to the hollow shelter, comb has many other functions. In winter when the bees cluster together for warmth, the densely packed, full honeycombs that surround them become insulating walls that help retain heat around the clustered colony. The queen lays eggs in the comb cells in a large internal area that functions as the brood nursery for the larval bees. That area is surrounded by the pip's pollen food, and further away, honey is stored for the mature bees.

The wax comb also facilitates communication throughout the entire hive. By vibrating the comb, the bees can easily communicate what is happening at one end of the hive to the other. The comb picks up the constant kinesthetic communication among the bees. Each bee touches and is touched by other bees hundreds of times throughout the day. This contact is a continual confirmation of "us," affirming the bee's awareness of its role as part of the whole and reinforcing the hive's unified consciousness.

The plastic comb conventional beekeepers often use in place of bee-built wax comb is too hard and brittle to function as a communication system. The bees say it is a frustration to try to vibrate it. Bees with plastic comb still communicate as best they can, but brittle foundation is not optimal for them and limits their perception of hive events. Plastic foundation makes it harder to have a true communicative harmony within the hive.

IN OUR OWN WORDS

The Unity communicates through the comb. Each bee movement is telegraphed throughout the hive by comb vibration and bees brushing by each other while performing tasks.

Movement within the hive conveys a harmonic vibration based on the speed and regularity of a movement. Tasks are known by their distinct movement "sounds." Thus, we understand that new comb is being built because the comb-building bees have a specific rhythmic vibration they contribute to the Unity. Comb-builder movements let everyone know that these bees are pressing and sculpting wax and grooming comb cells in a specific area. The pressing motion vibrates the wax comb with a different sound and sensation than when nectar is fanned into honey. Each task is progress, and the number of bees allotted to each task is known to all through the comb vibration. Thus, we are aware of births, pollen storage, nectar preparation, the queen's location, temperature within and around the nursery, moisture levels, the efficiency of propolis sealing, and the comb's structural integrity.

This ongoing kinesthetic noise is part of our song, and we sense it as a continuous communication. The sound and feel are embodied within the hive as a language. Comb that is structurally firm, yet pliable enough to vibrate with the different tasks, is important to each colony's perception of itself.

The vibration of multiple bee presences reverberates throughout the hive by way of the comb. Simultaneously, each of us is individually present, contributing our place within the hive's Unity, and each of us is also the whole of the hive. All of us together communicate to the Unity the hive's full, combined presence.

The queen's location is known by all. Her caregiver contingent announces itself by its sound and vibration. They move across the comb together as a solid platform. She is a stability upon the comb, distinct from the smaller motions of other tasks. The queen is the sun in our constellation and the central presence within the hive. Each bee, through the contact points of bee-to-bee and the ever-present comb vibration, knows where the queen and her contingent are at every moment, and that knowledge soothes us.

We replace wax where the honeycomb has accumulated old thoughts that no longer serve. There is no loss in removing and renewing what has gone before. The co-creative energy is always available for these places.

Every motion of a task tells the progress the Unity makes, letting us know how the hive is doing. Good comb vibration carries and reinforces the beloved, harmonious sense of right action. Industry is both progress and survival.

Constructing the Comb: We Build as One

In conventional beekeeping, each fixed frame within a hive has four rectangular sides and a flat sheet of plastic spread across the middle. The bees add wax directly onto the flat sheet, building it into three-dimensional combs.

Phil Chandler, author of *The Barefoot Beekeeper*, says, "Bees naturally build comb in deep, catenary curves (the shape made by a chain or rope suspended by its ends). But the use of preformed foundation inside rectangular frames forces bees to build comb according to *our* requirements, not theirs. Bees prefer to adjust the size of cells according to their needs."

I try to do things more like bees do in the wild. Instead of combs with fixed rectangular frames, I use hives that have a simple top bar—just a strip of wood—and let the bees build the comb below the bar in the way, size, and shape they like.

Usually they build a pretty straight comb, though sometimes it can be quite a maze, especially in vertical Warre hives. In Warres, the bees build to retain heat and scent, with little regard for straight lines. Their combs can be architecturally quite creative—more like the feral hive would build—unfettered by right angles and flat sheets.

A top-bar hive is long and horizontal, like a fallen log or sideways branch. The bees start at the top of the wooden bar and build their pristine white comb down in a hanging elliptical shape. If you pinned two pushpins to the top near the outside edges of the bar and then hung a thin gold chain between the pushpins, the ellipse of the chain is the shape the bees create. Sometimes the comb is wide and the curve is long; other times the comb is narrower, with a deeper pitch in the curve. Sometimes the bees will start two combs on either end of the same bar, and, miracle of miracles, the two combs will join perfectly in the middle. I like to think my husband and I are well aligned in our shared vision, but I will readily admit that if he started building on one end of a wall and I started on the other, we are just not unified enough to have the wall meet neatly in the middle. Yet the bees, without measuring tools or math skills, do this perfectly and precisely every time.

Recently I read a scientist's article about bees in which the author said that bees simply don't have a big enough brain to do all the things they do—and to do them so well—yet they still do all of these things. Science doesn't yet understand their unified intelligence.

This brings me to my tale of gymnastic bees.

I've been in and out of hives hundreds, maybe thousands, of times. Usually I have a reasonable expectation of what I'm going to see, but sometimes the bees surprise me completely.

I had just hived a swarm, and a week later, I thought I'd peek in to see how they were doing. I opened the hive and removed a few top bars. Lo and behold, I saw a secret acrobatic performance.

A few hundred bees were latched arm to leg, hanging down from one of the bars in an amber-colored veil. They were suspended there, about fifty bees across and twenty-five bees down, strung like jewels in an intricate necklace of sequenced layers. Each bee held with her foreleg the hind leg of the bee above, all of them in matching rows. I was dumbfounded. What, what, what were they doing?

They were using their own bodies to measure the placement of the comb. These draftsman bees would then convey a mental picture of the completed comb—including the height, length, width, and pattern—to the maidens who would construct it. The hive's shared consciousness is what allows the bees to use this system of measurement and construction to create the comb. Amazing.

IN OUR OWN WORDS

We have a consciousness that extends beyond one and into oneness.

When we hold our arms together in the open space, we are first building a template, spreading the measure of our bodies across the open space. It is as if we are standing upon the comb already built, as it exists in our perception. We see it there, and then we make it so.

The comb is already in our consciousness, and so we bring it to the present, as we first beheld it when the scouts saw it constructed within the space we chose as home. The thought is expressed in love of one another. Upon the fabric of love it is written.

When we hold this frame, we are moving between the work completed and the work as yet undone. These are simple steps, and our work becomes manifest. We bring forward what we imagine.

Herein we create a small part of the world with each one of us knowing naught more than the prescribed steps for one. But the one alone cannot bring even her part forward without the dreaming of the entire.

In the consciousness of the Unity, we build as one, both in space and time. We move freely between the place of imagining and the place of "it is so." The Unity moves between this natural fabric of time and matter.

Manifestation can be done, as humans sometimes do, using emotion as fuel, but this can bring many entanglements that disrupt the clarity. We can also enliven our imaginations by acting as one and in unity, which is simpler, cleaner. Thus, the world we imagine becomes the world in which we live.

May you all live as one and build your home around the heart of your nature.

Temperature: Enhancing Perception

When people discover bees living nearby, they often want them out—and fast. A handful of talented and compassionate beekeepers in our local group offer a bee-removal service to prevent the hives from being poisoned by pest-removal companies. My friend Susan helped with a bee removal for the first time in the summer of 2014.

The bees were in an old church. Two experienced beekeepers, Wes and Tel, began sawing open the church's gutter eaves to access the comb. Susan was assigned the job of tying the wax combs onto slim wooden bars, which she would set into a new hive box, lining up the comb like hangers in a coat closet.

This task sounds fairly straightforward, but its execution is not simple. A piece of fresh comb from deep inside a living hive is as flexible as a hot pizza slice. Both sides of the wobbly comb are completely covered with bees, sometimes several layers thick. And the bees are moving all the time, lickety-split. Even in the midst of an eviction, they industriously continue on with their hive work.

Susan tied a thin organic hemp string to the bar and ran it down the side of new comb, around the bottom, and up the other side, neatly tying a tiny knot to the bar to finish it off. The string hammocked each hunk of comb so it wouldn't fall out. Thus, each comb was affixed into a frame. She worked ever so gently with her fingers so no bees were accidentally squished. The task needed to be carried out barehanded, because one simply cannot tie tiny knots in loose gloves with a dozen bees on one's fingers! There is also risk involved: bees can sting when frightened by an inadvertent squeeze.

At the end of the day, when Susan told me of her experience, she emphasized two things: the astonishing gentleness of the bees—she had held fifty thousand bees in her hands that day—and the surprising warmth of the comb and of the bees themselves. "The comb and the bees were warm as melted butter," she said.

I've felt that same heat when I've moved swarms. Once a swarm settles onto a tree branch, all the bees light one atop the other until the clump of bees approaches the size of a football. That's when I show up. I don't need many tools to move a swarm; normally a cardboard box with a lid is all I need. I approach the swarm, place the box underneath, and give the branch a good shake. That shake makes the bees lose their grip, and in one fell swoop, most of them fall into the box. I put the box in my car and drive them home. Easy enough.

But sometimes the swarm settles onto a difficult place, like a chain link fence or a convoluted metal pipe that doesn't allow a quick shake. The first time I was confronted with this situation, I knew the best way would be to move the bees carefully with my bare hands. That way I could lift them off the fence a few at a time. With bare hands, I'd be able to feel if a bee was getting squeezed around the wire and could modify my hand position.

When I took off my gloves, all my senses were wide awake. What would the inside of the cluster feel like? Crawly? Stingy? I had no idea, but I felt nervous. I didn't know if I would get hurt or if I would hurt them. Both ideas scared me.

I asked the bees to show me how best to move them to a safer place. After a little prayer, I put my fingertips on the surface of the swarm. Ever so slowly and gently, I pressed my fingers into the clump. The bees opened like the Red Sea parting for Moses. I reached my hands in far enough to cup a few hundred bees and gently, gently lifted the warm cluster. Each time I reached into the swarm, I saw little bees benevolently looking up at me, and I felt the kindness we were sharing with each other. I marveled at the coherence of the bees in my hands. Such trust they had! Equally amazing was the heat inside the cluster. The bees surrounded and enclosed my entire hands with the most soothing warmth I'd ever felt. The heat of the hive then went deeper into my hands and enveloped my fingers, thumbs, and palms in a way that woke every nerve in my hands and opened all my senses.

I continued merging my hands into the cluster, tenderly carrying each handful of bees to their new home and then laying my hands on the edge of the hive so they could walk in and join their family. My hands tingled with warmth even after I was done, and I felt sated in body and soul.

Bees are masterful at calibrating temperatures within their hives because they need to have a specific temperature range for molding wax, processing food, and keeping the pips alive and well. With astounding accuracy, they keep their nursery at the precise temperature for pip development. They accomplish temperature and moisture stabilization in a variety of ways, all of them engineering marvels.

IN OUR OWN WORDS

Unlike ambient temperature, where close is good enough, within the hive, precise thresholds of heat must be reached and maintained to support the life processes. Comb is designed to allow for temperature constancy. We communicate with each other throughout the hive as we build comb. We understand the complexity of drafts and airflow. We build flat or curved comb to prevent cold or moisture from getting into the interior of the hive. Curving comb keeps wind and drafts away from the brood nest, directing heat into the interior and sending the cold draft back outside. Straight comb allows for air circulation, especially to cool the hive. We cool the hive by fanning en masse—a very enjoyable group activity, like singing around the campfire.

In hives that we design on our own, the air inside is always moving. In hives designed by humans, we question whether we can efficiently maintain appropriate temperatures or whether we have to do that ourselves with our body heat, which depletes us. It is better for us to live in hives that enable us to efficiently control the temperature and humidity fluctuations of the days and seasons.

When more heat is needed, some bees advance to the brood nest and generate heat by presence or by increasing the vibration. Heater bees rapidly transform food into heat energy. They shiver and create a vibratory agitation—a burst of heat that warms the brood nest. They can do this for short periods of time, but it does drain them. It is an expenditure of energy for the good of the hive. Drones, with their bigger body mass, are warm and readily contribute heat to the brood nest.

Pips raised at less-than-optimal temperatures are functional, but they lack an extra perception, a form of intelligence. Pips need to develop in a constancy of heat within a very critical range. Correct temperature sets in play a chemical process that brings about proper development that only partly occurs if the heat is too low. Pips can develop in a less-than-precise temperature range, but the chemical process of their development will be incomplete. Pips raised in variable temperatures are still functional, but as grown bees, they lack the higher intelligence that informs their senses.

The planet Earth doesn't operate alone. Earth experiences a combination of planetary influences in waxing and waning cycles, and we bees feel these influences. When a certain temperature is reached within the hive, the influence of the planet Venus is felt in the developing nervous and extrasensory systems of the bees.

These planetary cycles also affect plants. Bees with heightened sensitivity to the planetary influences are able to enhance the fertility of plants because we pollinate at exactly the right moment, approaching each plant at its peak nectar and pollen production time. Do you see, then, how temperature is deeply related to a bee's relationship with her plants?

Scent: Expanding Our Consciousness

Once spring begins, the colony's activity ramps up tremendously. I plunk myself down in front of the hive to see what's happening and am transported by the lovely scent that wafts out of the hive's front door. To help the field bees zero in on the entrance, a maiden will stand at the front door and release her delightful lemony, peanut-buttery perfume through the air. I find that scent hypnotic. It must be; I was once there for what I thought was fifteen minutes and came to find that over an hour had passed!

Everything the bees create has its own particular scent. The wax smell is due to age and what it last held. All kinds of floral essential oils are in the nectars and honeys. Pollen smells flowery before processing, and after processing, it smells yeasty. Propolis contains the resins of trees. Even the bees exude a smell. The queen's magnificent scent bonds the Unity together and tells the colony how potent her fertility is.

On a sunny spring day, one of our hives swarmed. I had placed my ear up to the hive earlier that morning and heard the hive's sound shifting: very little noise in the top box, some activity in the brood chamber, and lots of clatter in the bottom where the entrance is. An hour later, the swarm poured out and formed a giant buzzy cloud for ten minutes as I stood nearby, waiting to see where it would land.

Sometimes swarms fly off in a hurry and head to places I can't reach, like a cedar branch seventy feet high in the pasture, or they fly so fast we can't keep up running underneath them. Swarms who leave the farm are looking for their own brave new world where they can seed new colonies. Once I realize they are going where I can't follow, I stop in the field and watch their departure. I send love and strength to them, wishing them well as they set off to find a new home beyond the forest. But some swarms stick around. They land on low branches around the bee yard, places I can reach. I offer a new home in an empty hive, and if they accept it, I place them a distance from their original hive.

That spring day the swarm flew a hundred feet away and landed on the leg of a table in the garden. The bees clustered there, just beneath the top edge, and settled in, waiting for the scout bees to find them a new home.

I brought an empty hive to set on the table and placed a wooden shingle at the entrance to act as a walkway from the table's edge. Using a long feather from one of our turkeys (my favorite bee-moving tool), I gently scooped up a dozen bees and placed them at the entrance. With an air of adventure, they sauntered inside.

As the bees strolled through the empty hive, they communicated to the swarm what they saw. A few curious bees came up from the swarm to investigate the dark hive entrance. They went inside, and a few more followed. Then, without any bees coming out to tell the others, clusters of swarm bees began marching in.

By luck, some beekeeper friends were visiting. I called them and my husband to come see where the swarm had landed. The six of us sat in the grass a few feet away from the hive, watching them take possession of their new home. The swarm got smaller and smaller as the bees disappeared into the hive, until only a handful remained on the table leg.

We had yet to see the queen. The swarm cluster protects and hides the queen, and she was in the very bottom layer, which was now mostly uncovered. Knowing she was in that little pile of stragglers, I told everyone to watch for her because she was a fast mover, and any minute she would be dashing into the hive.

Suddenly a curious thing happened. The next contingent of bees on the ramp, about forty in all, slowed before they reached the door and spread themselves out in front of the entrance. They lifted their hind ends to mist the air with their Nasanov glands, infusing the air around us with a delightful, sweet, honeyed-peanut-butter scent. I imagined they were telling the queen, "Come this way. The entrance is up here."

The enchanting scent rolled lazily in the air as handfuls of bees slowly rose, dipped, and hovered. The air around us shimmered with dozens of magical floating bees. One bee lit on my shoulder, stared inquisitively into my eyes, and I smiled back at her. Other bees landed beguilingly on a shoulder or an arm or brushed against our hair. They didn't really go anywhere; they just rose up and suspended themselves in the air, drifting this way and that, mesmerizing us as they bobbed in place. One friend watched a bee land on her wrist and walk in a circle. Another said three bees landed in the grass in front of his ankle, and he was transfixed by how they interacted. We breathed their scent, and for a brief moment, we were hypnotized by bees.

I'm sure that's when the queen darted inside. While we were entranced with scent and the floating bees, the queen scooted past. Not one of us was looking at the ramp when she scurried by. I looked up in time to see the scent maidens tuck their bottoms down, fold up shop, and dash inside the hive. As soon as she was in, the bees quickly closed ranks, and in a hasty minute they were all inside. The bees had created such a clever diversion that we didn't catch on until the queen was safely inside, once again surrounded and protected by her court.

The scent the bees had infused throughout the air was unusual because it did more than imbue the air with a pleasant smell. It also gave direction, calling each bee's attention to something important: where the hive entrance was. And when we inhaled it, this scent entered our brains, causing a global sense of spaciousness, an expansion of perception so engaging and wondrous that it literally boggled our minds. We slipped into the silken folds of mindfulness, unencumbered by rationale. The scent brought us into the fascination of "bee time."

And who could turn down such an invitation? Certainly not me or my friends. In that brief, expansive moment, each of us relished a feeling of personal contact with a bee or two—a simple communication, a lovely connection with the friendly little creatures. Each of us accepted the invitation to stand inside the bee space with them. No, we didn't see the queen, but we gloried in the time spent in presence with the bees.

If you watch long enough at the entrance of the hive, you begin to see how the hive thinks and feels, how the bees display their philosophy. One afternoon I knelt on the roof deck where we have a few hives. I had been watching the bees coming and going for some time, hoping to recognize who was doing what as they flew in and out with multicolor pollens. Nonforaging bees ambled across the entrance, cleaning up pebbles of dropped pollen. A small contingent of guard bees sniffed everyone to be sure only this hive's bees gained entrance to the interior.

I noticed a foraging maiden land at the entrance and walk in. A guard bee hustled up to the forager and checked her out. They spoke via their antennae for a moment, and the foraging bee started again toward the door. The guard bee scooted around her, blocking her entrance to the hive. The forager took a sideways step to go around, but a second guard joined the first guard, also giving the forager the once over.

At first I thought she was an interloping bee from another hive come to test the strength of this hive's fortress, and the guards were preventing her from entering. Then I noticed her wings were ragged—a sign of an older bee who has flown so many flights to the fields and back that her wings had begun to wear out. The normally smooth edge was tattered, and though she could still fly, she probably didn't have many more foraging flights in her.

The guards' antennae flitted around the forager in a rapid and intense conversation. It wasn't her ragged wings that told them this bee was nearing her end time. Every bee's scent is an indicator of the bee's vitality, and this one wasn't passing muster.

After three tries at entering, the forager stopped. To back up the guards' opinion, another guard joined the line. Then the forager suddenly turned and briskly walked down the ramp, away from the hive. Knowing little at this point, and clearly intervening where I had no business, I scooped her up on a feather and deposited her back at the front entrance. She still had pollen on her hind legs, and I thought the guards might have missed that.

Without a moment's input from the guards, the forager turned on her heel and again moved down the ramp, out and away. I scooped her up one more time, but everyone did the same thing again, with the forager marching away. I followed her to the edge of the roof deck, knowing she'd have to fly somewhere and curious where she would go. She surprised me when, instead of flying, she simply stepped off the edge and fell, plummeting into the rosemary bushes below.

She'd been turned away because she had completed her usefulness to the hive, and she both knew and agreed with the guards' assessment. I thought of indigenous snow cultures putting elderly family members on ice floes when food is so scarce there wouldn't be enough to sustain the whole group through winter. How painful it must be to make that decision, and how even more painful it must be to accept it. And here was this little bee elder, not protesting at all, but simply accepting the decision the guards were empowered to make, for the good of the Unity.

IN OUR OWN WORDS

Light and scent create our map of the world. Through either or both we read life force and thus a map of the living world. At the moment each flower comes to its peak life force, the pollen and nectar are taken. Life force holds within it the spiritual energies that convey themselves to us, and we eat and breathe them. The scent inhaled and ingested to us is every bit as important as the food itself.

Nutrition comes from food and from the scent of a flower at peak readiness. That scent nurtures and heals us. These collections of nectar and pollen invigorate us, stirring the hive's consciousness and expanding our being.

Within our hive we build a great hall. Each hive builds its own hall and makes the interior of the hive a portal. Halls are created through the structure of the parallel combs and the hexagonal structures on them. The scent and sound within the hive are like a key within a keyhole. The way sound moves through the comb creates the keyhole sound. When the scent and sound are right, there is a transportation—like entering an elevator—that connects us with all other hives. We are all present in the experience of the other hives when we are in our halls. We feel all the joys and all the sufferings they do.

When we enter a new hive, we atomize the air with our perfume, which induces a quiescence in any who look our way. This provides a veil of obfuscation so those who gaze our way may look but will not see. Whereas the queen scents to communicate to us, our maidens scent to cloak us from outsiders.

Our scent induces a dreaminess—a forgetfulness even. It heightens the scent we send and calls attention not to us but to one's interior. Those who look at us are slowly drawn to note how our scent affects them, and in that pause, as attention is drawn away from us and toward the other, the veil allows us to move freely within the shadow of the scent.

Each maiden bears a scent that conveys her state of health. When a maiden reaches the end of her cycle, her scent changes, and we know her end time is coming. Sometimes we remove a bee early because we know her end is near. Once signaled, she will exit the hive and walk away. Thus, the hive doesn't become cluttered with those who have passed.

Bees remove themselves by not returning home. They leave on their own once their end time becomes known to them. Even in their passing, they work for the good of the hive.

Anything within the hive that diminishes the hive's internal scent decreases bee health. In a weak hive, the stench of disease drains our life force. Rather than remove the disease, we know to remove the bees. Despite human interventions, those bees, and therefore that hive, with rare exception, cannot achieve its fullest expression. The directive is to let the weakness die with the hive.

Some hives, if left alone, can remove the disease from their colony. Scent is the healer. This is why it is so important that people not interfere. We have our own medicine. Some hives are capable of using the propolis in the manner it was designed—to heal the hive itself. We drink in the scent of our medicinals, the plants that heal us, the encapsulated light that gives us strength. Inhaling, drinking, drawing the scents into us strengthens our whole bodies, and light moves through us where darkness had been. In this way, a hive uses scent to bring about its own healing.

Sound: Constant Communication with All That Is

Bees don't have ears. You'd imagine they do, wouldn't you? Buzzing is certainly one of the qualities we associate with bees. How curious then that sound, as we understand it, is not in their language. Yet if we go a little deeper and define sound as vibration, we'll realize that the bees have plenty to say through their songs that acknowledge, define, and celebrate their activities.

Bees sing songs—each song for a different purpose—and these songs are similar from hive to hive. The wailing dirge of a lost queen's death is filled with sadness and is so identifiable to a human ear that a beekeeper need not even open the hive; the keening alone says the queen is dead. The Territorial Song calls bees to defend their home against marauders, and the bellicose Song of Assertion is a rallying cry when a battle is about to ensue between hives. Singing songs brings them together in a shared purpose. Colonies also sing songs that have harmonies to keep them in good health.

The most powerful of the bees' songs, the Song of Increase, is about robust health, but the bees explained that it is also about everything else that's going on in the hive. Within that song is the sound of every appropriate action each bee undertakes to move the unity of the hive forward: the sound of the queen laying her eggs, the sound of maidens turning nectar into honey, the sound of house bees repairing and making new wax, the sound of drones singing to the newborns, the sound of house bees producing fermented pollen, the sound of babies coming to life in the nursery. Other sounds communicate if the ambient temperature in the hive needs to go up or down, the quality of the food in storage, and who needs help with tasks. These many sounds, expressing everything happening throughout the hive, compose the joyful Song of Increase.

When I first learned of the healing harmonies of bees' songs, I immediately thought that playing the sound of a strong hive to a weak hive might make a weak hive stronger. The bees of the weak hive would come into harmony with the sound of the strong hive, and all would be well. What I came to understand is that the song conveys not just the sound of a strong hive, but also a full-on map of the bees' home in this space and time. By being a part of the sound, a bee knows what tasks are occurring in the hive and where, and she can move to another area if she knows the bees there need help. For this reason, I feel it's not useful to play the sound of a strong hive to a weaker hive. The song of a strong hive describes entire areas that don't exist in the weak hive, and hearing it would lead to confusion in the weak hive.

IN OUR OWN WORDS

Our inner landscape is sound. We move between narrow paths and lanes and know the way by the resonance of sound as the waves move onto and off of every surface in the hive. Besides the information about the hive's activities moment to moment, the sound itself plays through the hive and describes the curves and contours of the internal shape of the hive. We see inside the hive by the shape of the sound, which is one more aspect of Unity. In all our activities, we live inside the sound. Sound is integral in our Unity.

Sound is the emanation of our vibration. Each hive has a unique sound signature that declares its health and vital force. The emanation of the sound reinforces the hive's state of health.

It is as important as our warmth and scent to the hive.

Each bee individually has a sound declaring its health. In harmony together, we set up a wave of sound that speaks (in a song) of hive health. The overlapping sounds harmonize on different frequencies that speak to each of the bee's organs. The song can raise or lower the vibration of vitality, enhancing or decreasing hive health. Like a song of rapture, it can raise the spiritual frequency of the hive, or like an observation or complaint of ill health, it can wear down the bee's vibration.

Sound can heal bees by directly imprinting the pattern of health into our organs and immune systems; lifting up the hive with a more vital vibration; and unifying us in the Song of Increase, our most joyful emanation.

Each hive's sound describes that hive's health and status. Everything to be known about that hive is described in the sound. But this is not merely a telegraph of current conditions. The sound is an emanation from each individual in the hive, wrapped around and through each aspect of the hive. The effect of the sound as it vibrates through the structure also communicates something larger to the world.

In the world, there are healing chords that reverberate like tuning forks. These tones express love, and they give, through their very structure, knowledge that continues the development and evolution of the world. Sound is the medium through which creation flows. The thought "I am," spoken, becomes being and matter.

Throughout the world, sound sings our home into being. Especially vital are places where there are many healthy, uplifting sounds—waterfalls, wind in trees, people using gamelan gongs and singing bowls, contented purring and laughter, the thundering hooves of a herd, chattering insects, and singing birds. Sound sings in tones and drums, as well as in caverns and natural places where confined sound magnifies itself. Humans sing spirituals, chant, whistle—all with a desire to heal, to impress the sound of their living and being upon the air.

Even the sound of a tornado or hurricane is a prayer, as it washes clean the land and returns it to the fresh medium of silence upon which the new can be written.

Thus, sound is a returning place, a way of retuning ourselves to our harmonious original thought. Within the vibratory structure of a sound is the path to the matrix of matter. Returning to our sacred and uplifting sound recalibrates us, heals us, and allows transformation and evolution to enter the intended higher place. The sound of the hive is healing. We sing the sound of our creation, a constant communication with all that is. The sound tells each bee what to do, where to be, how to engage. Each hive task is part of our sound. When all is right, the harmony is perfect.

Food made in such contentment nourishes our body and is balm for our soul. Honey is food made in prayer. Sugar syrup is too high pitched, like treacle. We eat it if we're hungry, but it makes our stomachs hurt, and it doesn't have the prayer in it. Sugar makes our singing weak and tinny.

In the hum of the hive, we see the hive entire. When we begin building, the imagination of our home already exists. We, in our industry, fill it in. When we visit a new home the first time, we enter and see our home hive already completed and filling the space. Our actions are simple: we see through our Unity's internal awareness, and from this we draw forth our imagination and express it into matter. We build ourselves from the inside out, from the thought to the matter.

In the unity of the hive, the flow of steady progress brings out the Song of Increase. A hive singing the Song of Increase is in full harmony. This is the highest state of being within the hive, the time when every bee is in right action and the hive is in full expression, when love flourishes within the hive. This is what all hives seek—to sing the Song of Increase. This song nourishes bees by encouraging an expansive presence in each bee. Thus, we move within the hive with an emanation larger than our bodies, rippling off our hairs, which gives the impression that each bee glows, incandesces, within the darkness of the hive.

Question: What would happen if a beekeeper introduced the sound of a healthy hive into a weaker hive? Would the weaker hive become stronger?

This has not been done before, and we are wary of quick solutions that do not address the underlying problems. The intention is correct, but the application has an error. What you intend by introducing the sound of good health is a right direction. The problem is that it contains too much specific information that will cause confusion in the weak hive's tasks.

The sound of the healthy hive contains the tuning of good health and cooperation, but it also contains the hive's comb map and expresses tasks that may not be appropriate for another hive. The sound of a hive tells that colony all that is happening within the hive. While the sound of the healthy hive could be helpful as a retuning, the structure of the healthy hive is also in that sound and would be overlaid onto the weaker hive. The expression of the stronger hive will not match the layout of the weaker hive. The vibration of the sound is confirmed by the movement of the comb communicating the hive's activities, saying, "Here the queen is laying eggs, and over there we are packing pollen. And here we are drawing down nectar, and here we are shoring up with propolis the weave of wax." These tasks would not match what the weak hive is doing in their hive.

A better sound would be taken from a new hive as it is setting up housekeeping—a freshly started swarm. This new colony is organized but still in the building state, so the creative energy is still being focused. They are drawing out the rooms from the projected imagination. Thus, a small hive may hear (by being immersed in) the direction in which a strong hive is going. It may encourage the bees to draw from within their own forces.

But this cannot replace good genetics. It can only help if the core strength is present, but not fully awakened. Good genetics are 100 percent necessary. A weak hive will eventually die off, and this is appropriate. For example, a secondary cast from a strong hive has good genetics, but may be hampered by small size. We need a volume of bees to build comb and feed everyone during the primary construction. Otherwise the queen will not have enough room to lay eggs to increase the population. The continuing smallness of the hive weighs on us. We seek a volume of sound around and through the Unity that speaks of our success, which encourages more activity.

The simplest solution is to add brood from a healthy hive and use the sound of a healthy hive as a brief support. Be watchful of the effects of the sound, and do not use it for too long. Too much would cause harm.

Do not introduce confusion. If the new sound riles, immediately remove it. Listen to the small hive while playing the sound of a young hive in its new building phase, but play the new sound from a distance away. If it creates stress, do not continue. If the smaller hive comes into harmony with it, let it continue for a day, possibly two days, but no longer. Then diminish it so they can take up the sound themselves.

This is rocky ground because it is an artificial stimulation, but it is appreciated that you seek to increase strength and vigor in the weaker hives. Though a hive is weak, that does not mean it is wrong. Weak hives also perform a service because they tell nature that something is awry and out of balance. These hives give service as markers of imbalance. Their weakness is their service.

The underlying issues in nature must be addressed to heal the bee family. Better actions are creating sanctuaries within the world—havens full of sound that involve nature in every aspect. This is where true healing happens. Hives placed in such havens become stronger. This is the best direction through which to bring strength back into the hives and continue the evolution of the world.

Propolis: The Envelope of Our Medicine

In January, after a few days of hard frost, I noticed that the bees who live in the outer wall of our farmhouse were particularly active. The air was still quite chilly, yet little clusters of bees excitedly scampered out their doorway and flew off into the cold sunshine. At first I thought they were doing cleansing flights, hopping outside long enough to defecate then swooping back inside to the hive's self-generated warmth.

But no, these bees flew off toward the trees, and when they returned, their hind legs were covered with bright yellow pollen. Where in the world did they find flowers blooming in the brittle cold of winter?

I walked all over my farm, looking for anything sprouting up from the frozen ground, but I just didn't see any plant that looked capable of providing pollen. So I positioned myself where I could see the bees exiting the hive. From underneath, I saw them silhouetted against the sky, backlit by the sun's low rays as they flew over the top of the roof to the front yard. I followed them around to the other side of the house and watched as they flew to a bushy, ten-foot-tall hazelnut tree. Long clusters of golden brown catkins in full bloom were the center of each bee's attention.

Surprisingly, some trees do bloom in winter—willow, witch hazel, hazelnut. The pollen from these trees is particularly healthful to the bees, and on any warmish sunny day, they will be all over the tree's flowers. Sap also flows from some trees in late winter. On our farm, cottonwood, spruce, and other evergreens contribute their sticky sap. The bees carry home winter sap, which they make into propolis, the source of the bees' medicine.

Propolis is a sticky, resinous substance of varied colors. The bees create it by mixing tree and plant resins, beeswax, pollens, essential oils from flowers, and their own enzymes and glandular substances. The chemical and biochemical properties of propolis vary according to the bees collecting it, the collection areas, the time of year, and even the time of day. The propolis itself is an alchemical blend of hundreds of biochemical processes, some known and some still unidentified.

Bees make and use propolis to fulfill a wide variety of the colony's needs. Bees build propolis canals on the walls and ceilings of their hives to move the condensation water that comes from both the high interior hive temperatures and the evaporation process that turns flower nectar into honey. This water is directed away from the combs and allowed to pool in certain places so bees can drink water without having to leave the hive.

Hive walls and surfaces are coated in thin skins of propolis. The propolis acts like weather stripping for cracks, keeping out drafts that could chill or dry out the brood and harm the pips. Many viruses, certain bacteria, and pests cannot survive in such a hot and humid environment.

Conversely, many other bacteria and yeast thrive in such a damp, dark environment and could endanger the colony if unchecked. Propolis is known to have powerful antimicrobial, antifungal, and antibacterial properties that counteract these dangers. The propolis releases strong resin and essential oil scents that kill off specific organisms harmful to bees. This same scent is also a healing balm and immune stimulant to the bees, who have little internal immune defenses. Propolis is, in effect, the external immune body of the bee; the propolis scent makes up the part of the bees' immune system that is actually on the outside of their body.

Dr. Peter Mansfield, the founder of Good Health Keeping in England, uses propolis in his medical practice and has this to say about it, as quoted in *Bee Propolis: Natural Healing from the Hive,* by James Fearnley:

> Some of its properties defy not just chemical analysis
> but the very principles of chemistry. The interior of the
> beehive is a remarkably clean place, sterile place—far
> more so than the surgical departments of most hospitals.
> Yet this is achieved not by dosing all the bees individually,
> nor by lining the entire honeycomb, but by an outer skin
> of propolis alone.

In the same book, Fearnley says, "The beehive is a symbol of how simpler substances derived from the lower order of the plant world are elevated and transformed by the bee into substances appropriate for a higher order of existence."

Bees paint their comb with propolis to sterilize the cells for the pips. They use propolis to stabilize the wax so that it is better able to carry the full weight of matured honey. Propolis is used at hive entrances to open or reduce them. The bees use propolis to create narrow tubes they pass through when arriving and leaving the hive entrance. These reduced entrances are easier for the guard bees to protect from intruders and also serve as decontamination chambers for bees who have been out in the world and may have brought something harmful back with them.

Most of the bees on our farm are adept at modifying their front entrance to fit their seasonal needs. In fall, when the days are variably warm or cold, I've seen the hive entrances modified daily. One hive in particular changes their entrance as often as once or twice a week, mostly in summer and fall. They built a scrim of propolis just inside the wider entrance, and from there, they can limit access to prevent robbing, increase or reduce ventilation and moisture, stave off cold winds, or fling the doors open to let the multitudes of foragers in and out during peak nectar flows.

Closed, a scrim has an opening only as wide as two bees—a common situation in winter. I've seen other hives seal the door nearly shut when yellow jackets tried to storm the entrance. Other times, the bees may create multiple openings within the scrim. If I suspect the hive needs better ventilation, as one of my hives informed me, I may make a small round opening on the back side. They leave it open when they want to move the inner air more efficiently, and they close it off when such air movement is no longer needed.

Propolis can be used to seal over anything the bees can't carry out of the hive. A friend was doing a hive inspection in early spring when she discovered a lump of propolis as big as her thumb in a lower corner. She cut out the lump and found inside a dried-up mouse body. Apparently the mouse thought the hive a warm and inviting place to build a nest for winter, so it set up housekeeping in the corner. The bees, of course, wanted nothing to do with a smelly mouse who would pee and poop in their immaculate home, so they sealed the mouse into a propolized sarcophagus, safely entombing and mummifying the intruder.

The bees tell me of the healing properties of the propolis and describe the aroma as a kind of vibratory signature. Each hive, in its unique location, crafts a distinctive type of propolis that may differ from the recipes of its neighbors. This finely crafted substance is the exact medicine needed to surmount any weakness of that singular hive in that particular area. I see the uniqueness of my local region in the fragrant scarlet propolis created by the bees on our farm. A few miles away and at a different elevation, a good friend's bees make butterscotch-colored propolis that has an entirely different scent.

On our farm we do our best to keep the environment vibrantly healthy for bees. Our intention is to give them many different ways to support their good health, including a wide variety of seasonal plantings of herbs and flowers, especially borage, flax, thyme, and other herbs, and let them choose their own medicines. At the edge of our garden irrigation, we keep a pile of seaweed that gets doused each summer morning at dawn to keep it from drying out. I often find bees lined up on the edge of the seaweed puddle, sipping seawater with their hollow tongues that work like straws. The seawater gives them access to the full range of trace minerals that bees might need. My friend Glenna puts a handful of oyster shell chips in her watering station, and she says bees like that, too.

I trust the bee teachings and the wisdom of treatment-free beekeeping. Nature wants strong genes in the gene pool, so I let weak hives die and have confidence that the stronger hives will survive, improving the gene pool as they do. To that end, I don't treat with chemicals, medications, or even natural products. Instead, our farm has many trees and plants that provide good nutrition and ingredients for them to make their own medicine: propolis.

Sealing and Healing the Hive

It's my belief that a diseased colony can rebuild strength through maintaining a sealed environment. How the bees keep the hive sealed is important because it ensures their access to all the benefits of the propolis, especially in the hive air.

A few years ago, we inadvertently brought a colony with deformed wing virus (DWV) into our bee yard. Mites lay their eggs in the cells of unborn pips, and when the baby bee is born, it comes out into the world covered with blood-sucking mites, which weaken the bee as it matures. These mites can be even more devastating when they carry DWV. The virus infects the pips and prevents their wings from forming correctly. Without wings, bees can't fly out to the flower fields. No wings, no foragers.

I had collected the colony from a barn owner who wanted them removed from the structure's wall. I didn't know they had mites, and during the transfer I placed much of their mite-infested brood comb in the new hive. Soon after this colony arrived on our farm, I saw bees at the front entrance with mites on their bodies and realized my error.

My first impulse was to worry about them, which wouldn't change anything and would increase stress for everyone. Instead, I prayed these bees would become the healthiest they could be. I blessed them with good thoughts and let them handle their own recovery. I put my thoughts to imagining the bees strong and healthy, as a vibrant colony capable of surmounting any odds and becoming robust again. I resisted the urge to open the hive and check on them, and instead monitored the entrance to gauge how they were faring. The sanctity of the hive is precious when bees are dealing with a disease process. These bees needed to seal their home with propolis so they could begin their healing.

They began their healing process by removing from the hive any mature bees with the telltale shriveled wings. Next they went into the cells and began taking out infected larvae. They were thorough and dedicated. When I saw them pulling out infested pips, I knew this colony was going to make it. The entire process took a few months, and I could tell the bees were improving week by week. At the end, the bees were mite-free, once again strong and robustly healthy, and they have continued on that way. Although being mite-free may seem desirable, bees do not need to be 100 percent mite-free to be in good health. They simply need to be capable of living in balance. I say this because I have found feral hives in both situations—with and without mites—yet vibrantly healthy because they arrived at their current situation on their own terms.

I believe bees who heal in this way are propagating an appropriate self-generated response to a disease, expressing their willingness and ability to deal with it. They clean up anything associated with the virus and eventually are either free of the mites or have arrived at a comfortable symbiotic relationship wherein they are not overrun or debilitated by a small quantity of mites.

Letting Nature take its course is a long and sometimes painful path for a beekeeper to walk. I believe this is how bees direct themselves toward self-preservation. They will survive if we can keep our fingers off them long enough to let them bring forth processes that are already present in their collective mind of their knowledge, so they can develop new natural systems.

IN OUR OWN WORDS

Propolis comes from the trees. Sap is the blood of the trees. We are, as you would call it, chemists—or more appropriately, alchemists—who create our own medicines. This is not a random mix. We know when we have enough of one and need more of another.

The resins release powerful scents into the air and all around us. We live inside the envelope of our medicine. Bees seek to live in hollow trees because the natural oils of the trees will be combined into the propolis recipe, which is aligned with the living system of bees and other beings. Propolis encourages the positive, life-supporting organisms and conditions, while discouraging those that are less conducive to good health in bees.

Humans have undervalued the power of scent and sound. Imagine your home as if you lived within the holy space of constant and ongoing healing, surrounded by scents and sounds that continually bring you to more of your self. You might ask, "Does this sound enrich me into expanding consciousness? Does what I breathe into my body bring health and wisdom?"

The structure of propolis has different keys that fill our damaged places and stimulate the limbic system to draw off the poison by wrapping it with a woven blanket of frequencies that deter, disable, or dissemble its course. The scents have frequencies that render the invading process to nil. The frequencies contained in propolis latch onto the imbalance and bring it again to equilibrium.

Health is calibrated inside a narrow frame. The stronger we are, the broader the frame. Weakened bees are precariously balanced, with little room for damage.

Just as we need many sources of flowers, we must also have a forest with wide ranges of trees. The trees freely share their life-blood; thus, they too are valued healers. We are as strong as our surroundings. Propolis is best made from many sources—some even quite small—so we can heal whatever ails us. The trees and flowers contribute to our good health, and we revel in their gifts and generosity. We revel, too, in the propolis air and our queen's glorious fragrance that surround us.

A wild hive is sealed with propolis where air enters or escapes. Breaking the seal opens the hive to outer influences that are hard to control. The propolis seal is meant to be nearly inviolable. When humans break the seal, we can repair it, but doing so draws energy away from progress. The flow of steady progress is a joy to us. When humans pull apart a hive to see what is happening inside, the healing, medicinal air is dissipated, diluting its beneficial effects. By sealing the air, we control our environment and retain the maximum benefits of this universal healing substance.

We encourage you to breathe the mixed scents of propolis. It will bring you to clear thinking—yet another aspect of good health.

Defense: A Challenge of Unification

Late in swarm season, about midsummer and well after most colonies had swarmed, I got a call to pick up a softball-sized swarm hanging in a tree. Because of their late start, this swarm had half as many bees as an early-season swarm. They had fewer bees to help out, which meant they would be hard-pressed to complete all their tasks by fall and be ready for winter. But I have a soft spot in my heart for swarms, big or small, so I brought them home and gave them their own hive.

All through the rest of July and into August, they toiled away, making fresh comb and filling it with brood, nectar, and pollen. Small as the colony was, it looked like they might, with a bit of luck and good weather, increase their population and put away enough for winter.

In late August I noticed a flurry of activity outside their hive and many bees in the air. My first thought was that this was the hatch of a population surge, and these were young bees doing their first flights. "Hurrah! With this many bees, they'll make it," I thought. But as I got closer, I saw a struggle at the front door, as guard bees fought to defend the entrance and keep outsiders from getting in and stealing their honey. The hive's guard bees were wrapped in leg-to-leg combat, stinging interlopers and fighting as hard as they could to keep the marauders from sneaking inside. The aggressor bees were flying wave after wave of advancing troops at the little hive's entrance.

I could tell the robber bees from the hive's own because of how they looked in the air. Foraging bees normally come back from the fields loaded with nectar. Their legs dangle forward in a tuck as they jockey for a landing spot at the hive entrance. The cloud of bees in front of the hive was empty of pollen. Their bodies were more elongated than those of foraging bees, and their legs hung out behind them. They were hoping to rob what they could from inside the young hive and then return with full bellies to their home hive.

Robbing, once started, is difficult to stop because the robbers have had a taste of what's inside the hive. In beekeeping, I'm usually quite hands-off, but I get very hands-on when it comes to robbing and protecting my hives. When we see robbing in our bee yard, we place homemade robber screens in front of the entrance of the hive being robbed. These screens complicate the entrance, which sometimes foils the robbers. We also put leafy branches in front of the entrance to make it more difficult for the robbers to find the entrance and easier for the guard bees to defend it.

The fact that the new hive was being robbed confirmed that it was weak. I put up the robber screen to give them a chance, and I waved a bunch of robbers away, but I knew this hive was not likely to survive. A day later they succumbed.

With this loss fresh in my mind, I asked why this happened—why this hive had to die and why other hives, who seem to start off strong, ended up failing, too. I wondered about mites and robbers and all the ways a perfectly good hive comes to an end.

IN OUR OWN WORDS

Mites are not the bad guys. Mites cull weakness out of the hive. Each divine bee has a gauntlet to run, something that asks or tests him or her. It may be the endurance of a long forage line, or the task of bearing up the measure of a comb length, or the frost initiating the call for extreme heat. Healthy bees meet each challenge, and the hive thrives.

When something is not right within the hive, we become distracted from the template of our tasks. Our sound changes—like when we can't find our queen, and a keynote of distraction enters our song. Instead of "All is well in the world," we begin to sing, "We are doing the best we can." It is a song that addresses a weakness that we know is present and is often a painful distraction. These distractions can be large or small. The song will always tell where the hive stands in relation to the challenge.

A challenge from without, like yellow jackets or robber bees attacking the colony, mobilizes the hive to defense, and a strong, sharp note presses out of them. This is the Territorial Song, a defensive wall of sound, which pushes out the entrance and lays claim to the hive's territory. The defensive song emanates from all the bees inside the hive, bolstering the defenders of the hive—the guard bees—who know the hive is in full consciousness of their mission to protect the hive.

Likewise, robber bees have a Song of Assertion. Always, the challenge is to cull out weakness, and robber bees are in service to the culling in the bee community. If won, the attack culls the weaker hive, and the stolen plunder from the weaker hive increases the stores of the robbing hive.

Before the attack, the defensive hive had tried continually to move itself to strength and harmony. If the hive is below par, the disturbances they experience—too few foragers, not enough comb being made, insufficient pollen—all draw attention away from the regularity of the hive's tasks. They respond with less solidarity and ultimately fail.

In most situations, the home hive has the advantage, and if they are able to maintain their cohesiveness, they easily throw off the challenge. Survival comes if the hive defends properly and the robbers withdraw.

This test is a challenge of unification. The timing is precise; the bees must respond in an instant and cannot be distracted. We need the ability to shift the efficacy of purpose to meet the challenge. This is a test of timing, punctuality, response to an alarm. When a challenge comes from within, the hive draws strength from stability.

At every moment, each bee has a single-minded purpose. An attack calls on our ability to shift, to know what needs to be done next. We depend upon the unified body always knowing what is needed in that moment.

Silent Robbing

When you spend time watching your hives and being involved with them, you learn each hive's unique personality, and you learn to recognize when something is off. I make sure to listen to my bees so that if they are having an off day, they can let me know in little, gentle ways short of a sting. One day in midsummer, a robbing incident between two of my hives made me especially watchful, and because of that, I was able to witness a remarkable event in my bee yard.

Sometimes robbing happens when the weather has worked against the bees, as when a drought dries up all the bloom or too much rain has depleted the flowers of nectar and pollen and shortened the season.

In such a situation, bees are stressed and may get desperate to increase their stores by robbing any hives they can find. The robbers may be honeybees, or they may be yellow jackets. In either case, the victimized hive may lose everything they've put aside for winter, and—even if they were once strong—they become weak or die off completely.

In this case, a hive I call the Sunshine Hive was being attacked by robber bees. I suspected a nearby hive, called the South Hive, was doing the robbing, as the amount of activity at the Sunshine Hive entrance matched precisely the activity at the South Hive entrance.

The Sunshine entrance was less than an inch wide. At this width, only one or two bees at a time could enter or leave through the door—a situation that gave the guard bees a better chance to keep the interlopers out. After watching in frustration as the South bees relentlessly rushed it all morning, I put a screen across the entire entrance and told all the bees that they had to stay where they were, in or out. I said that I'd open the screen at dusk to let everyone go back to their proper hive. Normally I wouldn't screen a hive in the hot summer sun because the interior of the hive would lose its ventilation and get too hot, but the Sunshine Hive was well shaded and cool.

I once heard that when robber bees are closed in together with bees of the hive they are robbing, they may become acclimated to the local bees and switch hives. When I opened the Sunshine Hive at dusk, some bees rushed out, but there was no flurry, and everything was much calmer than it had been. I kept an eye on this hive, and the next day, the robbing ceased.

In the meantime, I noticed something unusual at the South Hive. I'd spent plenty of time watching them through an observation window and was familiar with their normal busy pace. Now they seemed to be aimlessly wandering, as if they were passively looking for something. The behavior didn't make sense to me. I wondered if they might be queenless.

I rose at dawn the next day, before any honeybees were up, so I could figure out the situation. I looked through the South Hive's observation window. The bees were already active—earlier than the bees of any other hive. I watched them load up with honey, leave through the front door, and head over to the Sunshine Hive. They were flying with full bellies, their legs unmistakably dangling forward. The Sunshine guard bees seemed to recognize them and waved them right inside. It was robbing, in reverse. The South bees were pillaging their own stores and moving everything over to the Sunshine Hive.

Over the next day and a half, all the South bees migrated into the Sunshine Hive. My guess was something bad had happened to the South queen, and instead of the hive slowly dying off, they pitched their lot in with another successful hive and made a go of it.

After the South bees had relocated to the Sunshine Hive, my friend Tel came over to open the empty South Hive with me. We started pulling combs out and saw what had happened. The hive had a summer's worth of honey. In the brood chamber, however, we found not a single maiden egg. Plenty of drone brood was spread about, some of it already hatched, but no maidens. That told me the queen was gone.

Normally when a queen is injured or her fertility declines, the colony, pragmatically, replaces her. The brood chamber bees take on this task and form a queen cell right in the middle of the comb. They feed the tiny pip royal jelly, and the new pip becomes a queen when she hatches.

As expected, I found a queen cell right in the middle of a sheet of comb. The fully developed cell had a round hole in the bottom, which told me she'd grown to maturity and hatched out. Once hatched, she would have left the hive for a mating flight, but something must have happened. My best guess was that a bird ate her on her way back to the hive.

The new queen hadn't returned to the South Hive, and there were no young eggs ready to turn into a new queen. Knowing the hive's days were numbered, one of the maidens had stepped forward and volunteered to be a stand-in queen. Being unfertilized, she was able to lay only infertile eggs, which become drones. The nurse bees waited till the drone eggs were ready to hatch. Meanwhile, the South Hive had established a relationship with the Sunshine Hive and, on an appointed day, moved their honey and the rest of their family over there, leaving the nursery at their original hive unattended and empty.

I had been told a queenless hive that doesn't create a new queen simply dies off. At best, a drone-laying maiden might step forward and produce drone eggs, but after that last task, the entire hive predictably dwindles and dies.

Except in this case.

What do you do when you see behavior that flies in the face of common knowledge? I am in a beekeeping group run by my bee friend from Wales, David Heaf. I sent him the details, asking if he or anyone in the group had ever seen such a thing. David found this explanation of something known as "silent robbing" in the *Illustrated Encyclopaedia of Beekeeping,* by Roger Morse and Ted Hooper:

> Silent robbing occurs when robbing has reached a state
> where the two colonies are completely friendly with each
> other, and the robbers are going in and out of the hives
> amongst the rightful owners without being molested by
> guards. . . . The usual outcome of this situation is that
> the bees all finish up in one hive.

Unexpected compassion. That's what I saw.

IV

The Song of the World

The Communion of Bees and Flowers

Of the thirty to fifty thousand bees that a hive contains, about two-thirds are the foraging maidens. It is these foraging maidens who venture out into the world of wind, flowers, and light. Most of what is known about foraging bees has to do with pollen- and nectar-collection activities. After all, this looks like what they are doing as they visit thousands of flowers each day.

The bees are collecting pollen and nectar, but they are also doing more—so much more. The bees have shared with me a wondrous, epic story of mineral migration, planetary ethers, light layering, faeries and gnomes, and memory made into physical matter. Some of these messages from the bees are supported by new and ever-more mind-boggling research on honeybees and their extraordinary world.

We understand that bees collect nectar and pollinate flowers, but as the bees explain, they endeavor to collect only the very best nectar and pollen from the healthiest of flowers. This nectar and pollen must be collected at specific times of day and at specific developmental stages of the flowering plant. It is also the bees' task to see that these prime plant specimens are able to advance and thrive in accordance with the will and divine directive of the minerals and soil within the broad perception of the Unity.

Bees are exemplars of communication with all beings of Creation, as well as with their own kind. In this chapter, we follow the foraging maidens into their world to see how they communicate with other beings and what we can learn from them.

Enlivening the Green World: Libraries of Honey and Pollen

I grew up in a small New England town on country land with orchards, wetlands, pastures, hillsides, woodlands, and a pond. At a precociously early age, I knew the names, whereabouts, and bloom times of all the plants in our area. I noticed, too, how some plants, like the white trilliums that rose from the same root each spring, were fixed in their locations, while other plants, like asters and goldenrod, seemed to wander a bit. I watched yellow pond lilies thrive in our pond for many years, but then something changed and a surge of purple pickerelweed nearly replaced the pond lilies.

To this day I find great satisfaction in knowing the flora of my area and what blooms where and when. I have been influenced by the intriguing writings of philosopher-scientist Rudolf Steiner and the principles of biodynamic farming we use here on our land. Add to that my direct communication with bees, and it's probably no surprise I have a broad idea of what bees are doing when they pollinate flowers.

I am in full accord with Steiner's belief that the earth is populated with a wide variety of helpful nature spirits, including faerie beings, which I sense in the life force in different areas of our farm. It follows that I believe bees partner closely with the plant spirits or beings, to the point that we might say the bees are in communion with them.

The plants themselves are the emissaries, co-partners, and feet, if you will, for the mineral kingdom, the realm of the gnomes. According to Steiner, the mineral kingdom has a mission and purpose of its own: the minerals strive to create balance and harmony across the earth. They do their work slowly, over long periods of time, moving too subtly to be seen or felt by humans; but their movements are explicitly noticed by nature spirits, plants, and bees. Plant and tree roots draw up minerals and carry the news of each mineral's presence to the air spirits and the insects. The insects can move minerals by eating the plant, moving elsewhere, and dying, leaving their mineral-laden carcasses behind. Even though these actions seem small, over eons of time, a great deal of mineral movement is achieved by these means.

Plants flourish or fail as they wander across the landscape. They follow the directive of the mineral realms, which determine the soil components that feed each kind of plant. Wild blueberry bushes may thrive in one area for a century, roots happily ensconced in their desired blend of minerals. These particular minerals may eventually peter out and give rise to a different kind of soil. Then the blueberries will grow more slowly and bear less fruit. In their weakness, they become susceptible to diseases and die out, to be replaced by another plant who finds that combination of minerals and soil perfect. In great part, the minerals influence soil life in a particular area, and that determines the health of the plants and trees that grow there.

Some plants, like dandelions and comfrey, act as miners who burrow far into the ground where certain minerals live. Dandelions' deep taproots find calcium, iron, and other minerals and bring them to the surface via their leaves, flowers, and shallower roots. When the dandelions die back and decompose, the mineral-rich leaves and flowers are deposited on the surface of the soil, changing the mineral composition of that land.

I've seen this in our own yard. A dozen years ago the front yard was a riot of smiling yellow dandelion faces each spring. Over the years we have encouraged the dandelions to complete their task by letting them stay and not trying to remove or deter them. As these plants died, their bodies became mineral bundles in themselves, further feeding the soil and fully transforming the landscape over time. A dozen years later we see they have brought enough calcium to the surface that the soil there no longer needs their care, and the dandelions have diminished significantly. Apparently, now our north yard has need of their skills, so they have begun to move there.

Across the land, many plants participate in accessing and moving minerals. The job of each weed is to determine what is missing from the soils in its area and then try its best to create balance. That could mean mining minerals, creating vitamin-rich mulch, or significantly loosening up the soil structure so other plants can follow. Weeds work hard to accumulate the nutrients needed to make the soil more balanced.

What role do bees play in this mineral balancing and plant movement? The flowers need to know where to move next, and the bees convey that information and more. Every speck of pollen is a collection of knowledge about the plant, including where it came from and its connection to the mineral world around it. Bees perceive with their senses the unique chemical, nutrient, and mineral stamp of each plant. They understand this information as a language: the language of plants and minerals. This deep consciousness that passes between the bees and the flowers is a crucial aspect of pollination that far surpasses the simple fertilization we imagine it to be.

In the act of pollination, bees are performing two essential tasks. First, they move the floral pollen from flower to flower. These pollens contain information of past, present, and future mineral migration, and by spreading the pollens, the bees spread this knowledge among individual plants. Second, the bees impart an energetic stamp of awareness and approval onto each flower. This energy stamp, or transfer of consciousness between plant and bee, both begins and completes the plant's awareness of itself as a singular plant that is also part of the unity of plants in its family.

Every day each cluster of bees finds and connects with the etheric communication between plants. A group of bees works only with one type of plant during a foraging trip or perhaps a day's worth of foraging trips. A single bee can visit more than five hundred plants in one foraging excursion and can make more than thirty foraging trips a day. Flowers species are, for the most part, not mingled on pollination journeys. This allows for a more robust imprinting of the information that is to be shared among that plant family.

The bees know which plants are the strongest, and these are the ones they visit, moving the healthiest pollens across the landscape. I see this in my own gardens. When we first moved to our farm, I planted a garden in the front yard, but I didn't tend the soil well. Autumn-blooming sedums grow there and in other places around the farm, including some beds where the soil is near ideal for them. In autumn the bees visit all the sedums, but they visit the flowers in the better beds three times as often as the neglected beds. Even I can see the differences between the pale front yard plants and the more robust, deeply colored plants in the other beds.

As humans tend to do, I change my garden soil to suit the plants I want to grow, instead of sowing plants that will naturally grow in that spot. Only certain plants will find my soil perfect for their needs, so over the years my husband and I have loosened up the earth and added minerals that make the plant's food—soil—appropriate for the fruit, vegetables, and flowers we want to grow. If I want to grow squash and pumpkins, my claylike soil needs a lot of compost worked into it, because these plants have different needs than the blueberries that thrive in our natural soil.

By properly amending the soil, I can coax it into being more perfect, even for plants foreign to my area. But, alas, my heavy-handed method of amending the soil is not sustainable. A human is needed to add so many bags of very specific minerals and other materials to grow what I desire.

If left to nature, this task of amending would take longer but would be very efficient. If I ever stopped gardening, the land would revert to woodland and, ultimately, to forest in less than a decade. Without me, broad ferns would bloom in the shadow of big trees, swamps would nurture wetland willows and cattails, and sun-drenched flower fields would bloom across the land in the wake of wildfires and avalanches.

As bees gather nectar and pollen, they deeply sense the history of the land. Through the pollen and nectar, they understand the plants' intent to expand into their rightful places. Even in my beds, where the plants are well contained, I notice differences in how the bees treat them and the effects of those visits. The most robust plants get more bee visits, and the increased visitations seem to make the healthy plants healthier. The more vigorous plants multiply faster and are ready to expand into larger territory.

As a gardener, I have a responsibility to listen and respond to these plants, so I, too, am included in this conversation. Until I paid attention, I hadn't realized my edged beds restrained the plants' desire to expand, and I assume that was frustrating for them. Now that I understand the plants' communication, I comply with their desires. Plants this healthy need to expand. Taking a signal from their health, I give them more places where they can do what they intend. That's the best I can do as a human.

While I can add minerals to the soil and plant seeds, bees assist with plant migration in a deeper, more sacred way. In their awareness of a plant's history, they provide a mysterious and crucial service: *witnessing and acknowledging the essential presence of the plants.*

In physics we read about the capacity of the observer to change the observed; witnessing is a powerful, little-understood cosmic act or force. Each time the bees move from flower to flower, they bring their witnessing power to bear, and a jolt of life force is injected into the plant. The bees tell me that this interaction brings a "trembling delight to the plant beings and creates *joy* throughout nature."

As the bees began speaking to me on this topic, they mentioned that they work to "revitalize the ether." Since *ether* isn't a word I normally use, I looked up the meaning. I had a smile on my face as I read the Oxford English Dictionary definition: "a very rarefied and highly elastic substance formerly believed to permeate all space, including the interstices between the particles of matter, and to be the medium whose vibrations constituted light and other electromagnetic radiation."

IN OUR OWN WORDS

Pollination is so much more than cross-fertilization. Pollination is as much about enlivening the ether as it is about moving the reproductive forces. Both are important, and they happen as the pause between beats in music amplifies and describes the rhythm itself. Our task is to pollinate and to daily revitalize the ether. We are embedded into the task as it is embedded into us, and we have no singularity of it.

The trails we leave in the ether are webs of protection over the earth. Each distance we travel between hive and flower is enlivened. That's why we travel so far, because a closer distance would leave gaps or holes in the natural world. We enliven the green world. It is our charge.

Honey contains all the minerals needed for good health. Each mineral has a sound signature that creates harmony within the being who eats that honey. The designs of the flowers, their shape and structure, are part of the vibrational emanation. The world is so much vaster than humans imagine.

These songs are always being sung. These clusters of blooming color, effusing scent, and rising sound are all communications between life, constantly emerging in a cohesive, eternal collaboration that creates life and more life. All pollinators participate in some way. Honeybees, though, bring the spark of consciousness, unity, and love as they do their task and thus work presciently into the future to bring the heartful presence of love into the world.

We are the bearers of the energy stamp of Creation. The bee's exhalation infuses this energy stamp into the plant. The moistness is a trigger that wakes up the plant and announces the energy transfer, which the flower receives and acknowledges. Flowers are completed when the pollens are delivered. Each flower is fulfilled as it receives our encoded transfer. The flowers thus communicate through their pollination that they are in unison, bringing forth the same word.

Each colony has a significant role to play in the area where it lives. The hive needs to be acquainted with the local plants and the geological architecture through the seasons and through the decades and centuries, as the minerals reveal themselves in their slow migration across the land's surface. Bees are meant to be in their own areas. When humans move us, that compendium of local knowledge is shattered and lost.

The mineral story is told by the flowers, and then it is communicated across the landscape by the trails of bees in flight. It is ultimately held in our libraries of honey and pollen. The ordering within the libraries is chronological. The flow of seasons ripples across the comb and down the long sheet, much like a calendar. The flowers poured themselves into each drop of nectar. Thus, we know the minerals they shared and we stored.

During the winter vision time, when we eat the honeys made during different times of year, we are memory keepers of the flowers. These honey stories tell the history of the plants and the land's geological shifts. They contain the directive for the future of the plants and minerals, communicating where they are heading. In this way, bees and the winter-dormant plants remain vibrant to each other and keep alive the energetic unity between bee and plant.

Imagine you found a book along a trail that says, "Herein lies our history—the lands from which we have come, the places we have stayed along our way, and what each time in those places has done with us." The book describes generations of great expansion, the advancement into new areas, how the plant family flourished. And there are stories of great difficulty, of shifting weather and dearth of needed minerals. They tell the paucity of the days, the struggle to thrive, and sometimes a retreat when the minerals gave out in that area and the plants moved elsewhere. Mingled in the nectar and pollen we consume is our own contribution to the history of the plants.

We read this history from the groupings of nectars. They tell us how whole seasons unfolded, the diverse plants in sequence, and their relationships to each other in the mineral blends and mineral fields. We are particularly attuned to awareness of mineral presences throughout the land.

We don't read the maps of minerals like humans pore over maps to plan a trip, choosing on a whim where they'll go next. The knowledge in these historical and geological maps allows us to be present to a flow of information that already contains the route the landscape will take. We read and know in the same instant, having an emerging awareness of mineral knowledge, in its own slow movement, that delineates a path already laid in and through time.

The mineral map can be altered by subtle geological shifts as the land moves, opening and closing veins and mingling ores. Wind and birds spread seeds; ferns and horsetail arrive; mosses and primitive plants expand to cover and protect the exposed land, holding moisture so the reclamation plants can establish. Thus, new lands open.

Humans, however, greatly alter terrains, and these sudden shifts require the plant community to give much attention to bringing that land into balance again. Whole plant families rush in to disturbed areas to begin balancing what has been undone. When the land becomes too barren, greater forces become present and move minerals in a far larger manner by letting the seawaters wash over the lands. Thus, minerals are replenished in broad areas so we can begin anew.

Monocultures are deserts. They come from mistaken notions of the purpose of plants. Monocultures thwart the progress of mineral knowledge between plant families and the progress and transfer of mineral migration. Instead, altered mineral combinations are dumped into areas, and weed suppression begins. Although humans decide a certain kind of plant should grow there, little attention is given to the many minerals those plants most desire so they can thrive. Neglecting their mineral needs makes the plants weak.

We try to help by conveying this information to the flowers, but there are few nearby tracts of land for the necessary healing weeds to volunteer from. Areas of low diversity are an anathema to the plant and bee world. There is little to communicate other than lack and loss. The seasonal interrelations are missing, and the plants and bees dwindle and perish.

Rather than bearing life into the future, humankind has taken it upon themselves to harm and kill, thinking they are alone in the world and that life doesn't matter. When humans think themselves alone and kill without regret, they create a world where they are alone—a world bereft, a dying world. Where none survive, nature will renew, will start again, but without humans. Nature will knead them back into the dough and create something new.

Plant Fertility: Spiritual Nutrition

Three seasons of the year, our farm is in full bloom. We start with the earliest blooms of willow, hazelnut, and witch hazel in late winter and carry through to the last hurrah of goldenrod, asters, borage, and sedums—flowers stopped only by the kiss of a heavy frost. All the time in between, we have pollen and nectar sources galore.

Our bees visit all the flowers, yet not in the sequence they open. I sometimes wonder why our bees aren't all over some blossoms—ones I know they like—that seem ravishingly perfect for harvest. Masses of fragrant blue rosemary flowers or precious apple blossoms will sometimes be in bloom for days before our bees deign to pay a visit. Apparently bees see nuances in flowers that are beyond my own senses.

The bees describe plants as having a quality that I had been unaware of. They perceive what they call a *wafer* that exists in the plant's peripheral sphere. They conveyed to me the image of a flower existing in two places simultaneously. First, it exists as a physical object, standing in a field, visible and defined as you and I see it, in a state of wholeness, independent and complete. The second manifestation of the plant is the wafer that floats just beyond the outer edges of the plant; this is the medium through which the plant communicates to the bee its readiness to share its fertility.

IN OUR OWN WORDS

The wafer is a packet of information, like a scent—invisible but potent with knowledge of the plant, its health, and its readiness to mate. The wafers carry much information about the plant as it relates to the bee—specifically, about the readiness of the plant's fertility. Forager bees take in this extra level of communication. When near the plants, healthy bees sense wafers in the air that indicate the presence of peak fertility. Intelligent bees know to visit those plants at those peak times.

These wafers are not on the wind; they are around and above the plants. The wafers' unique effects emanate from the planet Venus, marrying heaven with earth. Even though the planet Venus exerts its influence on the plant's fertility, full fertility doesn't happen all at once. Sometimes plants cycle through a few days before their fertility peaks. Thus, some plants come into their fertility quickly, and other nearby plants come into fertility more slowly.

When the wafers alight on or near a plant, the plant, in acknowledgment of the wafer's presence near the bee, initiates the process of heightened fertility. The plant emits spore-like ampules—tiny multisided molecules—and jets of scent spores slowly explode out of the plant.

Bees with good sensory aptitude readily perceive these signals. The sensing is a "see smell" and informs the bees that these blossoms are ready for pollination. Nectar and pollen gathered at peak fertility have more life force and provide more spiritual nutrition for bees and for those who eat the honey gathered from those blossoms.

Bees who, as pips, were raised in an appropriate gestational heat are stronger and more perceptive; thus, they have the ability to gather the food with the most life force and to fertilize plants at the optimal time in the plant's cycle. When the colony's food has a tremendous amount of spiritual life force, the nourishment available to the bees is superior and full of energy-giving fuel.

Flight and Light: Emanating Life Force

It's commonly thought that bees seek familiar colors and recognizable flower shapes in the field, but I know they see more. I've learned that they see light differently than I do, with a wider array of colors than my human eyes can access. They also see the life force in and around every living thing. They go out each morning, seeking what they love, hoping to find wondrous nectars and pollens that will nourish them and their family. How joyous to have such a relationship with the earth and plants.

I often think myself a terrible farmer because I find it impossible to pull out all my weeds. Bees love the wild chicory and thistle flowers as much as the vegetables I purposely plant. The bees have taught me to see not just the functional "I can eat this" aspect of gardens, but also the life force that comes from everything in that garden, even the weeds. Many times I've dug up frisky chickweed, rambunctious vetch, and prolific lambs quarters like a good gardener, but I can't help noticing how filled with vitality they are. Instead of tossing these brilliant jewels on the compost pile, I've gone off and replanted them somewhere they can bloom to their heart's content and commune with the bees.

This morning I watched a bee hover in front of an exquisite sunflower. She gracefully landed on a point midway around the ring of tiny, spiraled flowers that make up the sunflower's face. In her momentary pause, she considered the ideal place to begin; as soon as she found it, she rhythmically collected the nectar from each floret in sequence, like walking a labyrinth. In the squash patch, I lifted leaves, looking for treasures ready for picking and found dozens of giant squash flowers filled with exuberant tumbles of golden bees, so covered with bright yellow pollen they looked like precious gems. The joy of their vigorous connection with the flowers made me laugh out loud. Oh, to be a bee in a field of flowers!

IN OUR OWN WORDS

Within the spiritual realm exist principles that bring forth effects far beyond logic. That doesn't mean these effects don't exist in the physical world; they just follow principles from another of the many realities that coexist in the natural world.

Such is the case with bees, flight, and light. Although logic describes bee wings as a surface area that displaces enough air to cause lift, bees fly using the concurrent system of levitation that allows the lift of elevation, movement through the air, and descent.

Bees also have a unique relationship with light. Their senses are keen to the many qualities light contains. They see and understand light differently from humans.

Light is laid in layers. The closer to the ground, the more descriptive it is of how it fills the space between life forms. It lies upon the surface of each life form and reflects an excitatory vibration describing that form's sheath. Beyond color, shape, texture, and reflectivity, light also conveys, in another spectrum, the life force of the form. Plants, animals, ores, and elements convey their life force by engaging with the light through a wave that surrounds them. The plant, for example, exerts an emanation that flows out beyond the plant's surface and interacts with the air around it.

Life forms fulfilling their roles embody a joyful assertion of presence and functional productivity, of participation in life. This assertion emanates into the atmosphere surrounding the life form and shimmers the air around it.

Bees rely on seeing a plant's life force emanation to know when the plant is ready to pollinate and gather nectar from, to know when a plant is most suited to our co-creative attentions. The emanation is a visible signal of the plant's success in fulfilling its role and apt progression through its life stages. Bees gather pollen from these plants. Thus, we ensure that the most light-filled plants within our purview are pollinated and carry forth their seed to the next generation.

Life is inherently exuberant.

Laying the Grid: Our Prayer for the Land

Whenever there is a birth, death, or marriage,
one must go and tell the bees.
OLD FARMER WISDOM

The bees had spoken to me about both light and flight. But one May weekend they made a request that showed me how they maintain a helpful relationship with both light and flight to intimately intertwine our and their activities with the good of the land.

Friends, neighbors, and our farm interns had helped us gather tractors, backhoes, and cords of wood to create a fertile, water-saving planting area for our fruit trees—a permaculture method called *hugelkultur.* Our crew of volunteers worked alongside us as we dug two ditches, each a hundred feet long, four feet wide, and three feet deep. We backfilled the ditches with seven cords of logs and brush and covered them with soil to a height of three feet. There we then planted our young trees. Over the next two decades, the wood will rot and create underground fungal growth that will feed the soil and enhance its capacity to hold water, thus keeping the young trees fed and watered.

The following day we hosted a farm tour for thirty-five people. We walked all over the farm, talking about the animals we raise, our gardens and orchards, and our approach to working with the land.

On Friday night, before all this weekend activity had even begun, the bees had told us that on Monday, we should plan to take the day off. They said all the activity from moving earth and having large groups of people walking around the farm would disturb the safety grid they maintain over our land. They asked us to do nothing on the land all day Monday so they could mend and restore the grid over the land. After the busy weekend, we were keen to accommodate their request and take a rare, but much needed, rest ourselves.

IN OUR OWN WORDS

When the sun comes through our wings and pours prismatic colors all around, we experience the transformative power of light.

The paths we fly above the earth create a cartography of flight lines, with each hive an anchoring point. The lines from each hive go out and come back, out and back—each hive a center point of the lines. Each neighborhood is covered by lines that weave over and through the entire area. These flight lines provide a protective web over the earth.

Bees are beings of light, and pollen is light. When we carry pollen back, we have a relationship with light as it goes through our wings. We fly and lay the light path above the earth, like constructing constellations. These paths, like garlands around the hives, protect the hive.

Our wings also have a relationship with silica. The light coming through our wings re-creates the crystalline silica pattern. As we fly, we bring silica into the atmosphere and lay it down on earth's surface. This activates the silica forces and draws plants upward.

We jubilantly fly around the hive and take great delight coming back to the hive with our gift. Inside the hive we are storing the light.

Every time we fly, purpose is laid upon the air. Our journeys are not merely ways to get from one place to another. Our paths in the ether enliven the air, making communications between the earth and the heavens more fluent. Our flight paths exist within the air far beyond the time we fly them. These flight paths are like the bars upon which songs are sung. They are the underlying structure of the earth-heaven songs.

Our flight paths exist much like telegraph wires. The elementals, the planetary influences, the guiding matrix—all have access, and communication flows. It isn't just information, because that can be shared many ways. Our paths create a convocation of spiritually adept runways upon which knowledge can be shared. Our paths are greased rails and, as such, are already blessed avenues for the intermingling of knowledge. The paths, too, are protected as vectors. The more they are used, the more powerful they become. As bees lay new paths and revisit old ones, they connect the places where plants thrive. Visits from bees nourish the spiritual qualities present in the plants, exchanging communications among spirit beings who live within and around the plants.

When we build our protective grid over the land, we also incorporate the heart space energy of those who live upon that land. These protective grids overlap and entwine with other grids, energetically connecting and reinforcing our prayer for the land.

We acknowledge the presence of humans on the land, and each, too, is woven into the fabric as an anchor point. A human with a force of will oriented toward bringing out the goodness of the land and all who dwell there is an adjunctive support to the work of the bees. A person who listens to the land, who observes and considers forward motion, who provides the needed and requested materials and labor for the development of the land, is known as reliable and is woven into the grid. In such a way, we recognize and incorporate life-enhancing and life-enlivening poles of being that come from industrious and vivacious adults, children, and central animals who contribute to the land's development and progression. These poles of being anchor and stabilize the grid. Thus, it is important when any being in the grid leaves, or a new being arrives, that one "goes and tells the bees."

While we do, on our own, notice and accommodate changes, telling the bees provides us with needed information to make timely adjustments in the grid. Our work as protectors of the land is respected when any major changes on the land are communicated to us, as sometimes earth changes may alter source points of earth energies that we then undertake to repair and strengthen. Telling us of these changes brings us into a deeper and more communicative partnership. This act both encourages and tempers the human will, so our wills align with the good of the land—our shared home.

Pollen, Field Cake, and the
Microbiome: Spiritual Nourishment

One of my favorite things to do is watch the bees bring pollen into the hive. All day long bees land on the front entrance, legs laden with firm balls of flower pollen. The hairs on the bee's body are charged with static electricity, and when bees gather the flower's nectar, little grains of pollen stick on the bee's body. After getting a good dusting, the bee tidies herself up by sweeping her front legs over her head and upper body, pushing the pollen to her middle legs, and sweeping until all the pollen is gathered neatly on her back legs. Her hind legs have stiff, spikelike hairs into which she presses the ball of pollen. If the pollen isn't sticky enough to make a clump, she'll add a dab of nectar to hold it together. When she has gathered enough, she flies home.

The pollen color is determined by which flower it came from and may have no relationship to the flower's color. Shades of yellow pollen come from maple, apple, cherry, peach, willow, mustard, clover, dandelion, blueberry, cucumber, melon, sunflower, pumpkin, and goldenrod. Gray pollen comes from blue borage, raspberry, and black-berry flowers. Pear and marigold pollens are orange. Malva is bright purple. Fireweed has a fuchsia flower, but the pollen is a distinctive royal blue. Phacelia flowers are pale lavender, yet the pollen is dark blue. One area of our garden has giant red poppies that give unusual but easily identified black pollen.

If you are a beekeeper and want to make sure your bees have plenty of pollen, you may want to plant heavy pollen producers. Borage blooms from early summer to frost. Goldenrod is a strong fall plant, blooming at a time when other blooms may be scarce. Echium bushes put out dense clusters of long-blooming flowers. Salvia is very easy to grow and has a long bloom. Exquisitely spiraled purple phacelia flowers are prolific in the production of both pollen and nectar. My website, Spirit Bee, has good lists of bee-foraging plants (see Resources).

The pollen is used as pip food. It is about one-fourth protein, and the rest is a mix of vitamins and minerals, fat, starch, and a little nectar to keep it bound. It also contains amino acids, lipids, and beneficial bacteria. The young nurse bees also eat some of the pollen, which helps their wax glands develop, preparing them for their next job as comb builders.

Because pollen has a dense cell wall that is not easily digested by the baby bees, pollen must be processed before it can be fed to the pips. The first step is to ferment it. Bees begin the fermentation by adding a bit of glandular secretions to the pollen as they pack it into the cells; they then seal the cell with a dab of honey. Microorganisms do the fermenting, and after a few weeks, the pollen becomes bee bread—or field cake, as the bees call it. Significantly different from raw pollen, the fermented pollen has many more vitamins and a lower pH. Studies have demonstrated that bees who were fed bee bread lived much longer than bees fed regular pollen, giving credence to the idea that the hive's microflora is more important than we had imagined.

A wild-bred queen, who has mated with a dozen or more drones, as feral bees do, brings plenty of genetic diversity into the hive. In such diverse colonies, each daddy-family of bees will have skills they are particularl good at. As we explored in chapter II, genetic diversity is important for the colony's health, because it gives the bee family a wide variety of traits it might need to survive and thrive in challenging situations.

Genetic diversity can impact biotic diversity as well. Internally, some bee families within a colony may carry the capacity for a strong microbiome, or the collection of microorganisms present in the bees. This biotic diversity could provide the colony with nutritional or protective advantages, as one small group of bees within the colony shares its beneficial bacteria with the larger community.

A study by Heather Mattila, et al (2012), found that colonies with more genetically diverse populations—feral bees—have 40 percent more active bacteria species in their guts than colonies from a single genetic line. Bees bred in large bee-breeding operations are generally from single or narrow genetic lines and are less likely to have such diverse gut bacteria. Buying bees from big bee breeders is how most conventional beekeepers acquire bees, so it follows that these conventionally kept colonies might not have the microbiomes necessary to be as robustly healthy as many feral colonies. Genetic biodiversity gives feral, swarming, wild-mated bees the gut bacteria they need to have a strong chance of surviving in challenging conditions.

The study also contained this nugget: "Because of the way that bee bread is inoculated, matured, and distributed, its microbial community acts as an extended gut for the colony, and the benefits of its activity are shared amongst all colony members."

Years ago I met Laurie Herboldsheimer, author of *The Complete Idiot's Guide to Beekeeping,* at a conference for treatment-free beekeeping. She described how microorganisms are responsible for fermenting the pollen which is then fed to the pips to make them healthy. When a beekeeper puts antibiotics into a hive to kill pathogens, it may have the unintended effect of also killing some very important bacteria that make bee bread so nutritious for young bees. Who knows what deficits would come of that? I was tickled to hear her say this, because the bees themselves had just begun explaining this idea to me, and hearing it from a scientific perspective deepened my understanding of the significant role of microorganisms in the bee gut.

The hive is full of bacteria, yeasts, enzymes, and more. The hive environment is clean, and yet it teems with life. A healthy hive has achieved a marvelous balance of life forms that coexist and, together, build health. Give bees nutritious forage, let them direct their home security with minimal interference, pray for favorable weather patterns, and bees are likely to do fine.

IN OUR OWN WORDS

Bee bread we call field cake. Field cake is made from pollen mixed with digestive enzymes that live inside the house bees. Eating the enriched field cake calibrates the pip's digestive system to match that of the hive. Pollen made from a familial bee is far superior to humanmade pollen cakes. It is not just fuel; within the cake are families of helpful bacteria. Eating these bacteria brings our pips into synchrony with their sisters and brothers.

These bacteria and yeasts are part of the generative family line, like the DNA of the hive. These related components replicate in all successive bees, promoting family. Field cake is the medium for the hive's internal flora and is how beneficial bacteria and yeasts continue regenerating their life forces through bees, symbiotically carrying all of them forward in time. Life begets more life.

Bees are place sensitive, and the pollen they eat tells them of the place where they live. There is a certain joy that comes when a bee has eaten pollen as a pip and then goes out and finds that flower's pollen as an adult. That helps the bee know that this pollen is good for the hive.

Field cake is a layered sentient blend of the light, scents, and tastes of the field. Full of spiritual nutrition and nourishment, it is food for body and soul. Field cake introduces the beneficial microbial and bacterial community to our baby bees, connecting each little bee to all bees before them. The pollens are mixed because that expands our palate and encourages exploration in mature bees. Field cake encourages us to find foods that taste like it when we go out to harvest.

The preparation of field cake preserves the internal organisms that help us absorb nutrients, guide immune system responses, and maintain good health.

The virtuous internal community has many layers. Each family of bacteria has a different role within the bee's body. Some bacteria break down the story of the cake so it is accessible to the bee's system, sorting and explaining the landscape and conveying a generational explanation of how bees historically relate to the spirit and substance of the plants.

When human medicine is applied to hives, it removes bacteria. This causes us to lose portions of the plant's relationship to our bodies, and we cannot absorb certain minerals. The bee suffers, though it may survive from this omission. Rather than protecting bees, the medicine causes a loss. True medicine would enhance functions, bringing about a deeper harmony within the system.

This is the purpose of the field cake: to strengthen and nourish all aspects of nutrition and build knowledge of symbiotic familial relationships. Human medicine is often medicine of subtraction, whereas nature's medicine works by addition, seeking to build strength and harmony within and between sentient awareness. Excellent field cake stimulates many positive responses and creates a splendid, health-filled, expansively knowledgeable bee capable of continually evolving to its fullest expression.

When treated by subtractive medicine, some portions of the microbial community die off. In a healthy world, specific microbes and bacterial communities interact with their partners to create a beneficial interrelated wholeness—the building blocks of life, where land, insects, and plants speak with each other. They need to speak with each other in love, directing and encouraging symbiosis, and constantly questioning and acknowledging what enters the bee:

Yes, this is good.
We need more of this. Not too much of that.
No, that has no function here.
Yes, this creates a desirable present and sets an evolutionary state.
Yes, in this moment we become expansive
and sing the Song of Increase.

Winter Dreamtime:
The Memory of the Sun's Light

In late fall I spend quiet time with the bees, sitting at the entrance to their hive as they finish foraging the last of the fall flowers. I hope each colony has enough honey put aside to ease through winter and that they have enough bee bodies to keep everyone warm in the cluster. Because of the slowdown in activity, some of the winter maidens exceed their forty-five-day life expectancy and live through the whole winter season. The last of the drones have already departed. They have no function in the hive in winter because mating with virgin queens won't happen again until spring. So, alas, out the door and into the cold they go, purged by the maidens. The queen won't lay more drone eggs until spring. I make sure all my hives are secured with tie-downs so the rare but occasional strong winter winds won't blow them over.

As winter progresses and fewer bees venture outdoors, I have to put my ear up to the hive to hear their presence—the constant low, contented thrum that carries them through the cold to spring. Here in the Pacific Northwest, it gets cold enough that our bees go in and out of torpor throughout the winter. Not a true hibernation, torpor is a metabolic slowdown that halts just this side of hibernation. If it gets cold enough, the colony stops all movement and sound. Anyone opening the hive at this time would think all the bees had died.

In winter they stay warm by bundling on top of each other like blankets. The outermost bees keep the bees underneath them warm. When the blanket bees get too cold, they burrow back into the cluster to warm up. Once a bee has completed her task of providing warmth to nearby bees, she will ingest a drop of honey from the winter stores, refueling what energy she burned off as a blanket bee. This is why it's best to have plenty of bees going in to winter. In a large hive, the blanket rotation is less frequent for each bee, and as a result, they eat less honey than a smaller, overworked hive would. A large hive can generate heat more quickly and with less effort.

At the end of summer I may notice a hive—usually a young, first-year hive—that doesn't have enough honey to make it through the winter. This sometimes happens with swarms that get a late start. If I've got an extra bar of honeycomb on hand, I may add it to the smaller hive. Besides being winter food storage, a densely full honeycomb stores heat remarkably well, providing substantial insulation from the cold, which helps the colony control heat within the cluster with less effort.

With small fall hives, I admit I have done more fiddling than I normally do. Besides supplementing their stores with extra honey, I sometimes ask a small hive if they would consider a merger with a larger hive. Most hives say they want to make a go of it on their own, and I let them do that, even with the risk of failure.

A handful of times a small hive has agreed to a merger. Typically, only one queen survives that merger, though it's not uncommon for a hive to allow the lesser queen to continue to live in the nether regions of the hive. This goes against common belief, but as I've mentioned, I've seen it happen in my own hives a few times now. It's always a surprise.

In a fall merger, the older bees will eventually die off, but the temporary boost in maidens greatly increases the hive's ability to finish the season more strongly. I've heard it said that doubling the number of bees in a hive brings about three times as much production: $1 + 1 = 3$. Larger hives are more efficient and have plenty of hands on deck to get all the work done before the cold weather settles in.

Every winter new beekeepers wonder, "Can I look at my bees in winter?" You could, but it would be better for the bees if you leave them be. You won't see much activity this time of year. On days that have a bit of sun, you may see some bees venture out of the hive for quick cleansing runs. They fly out fifteen to thirty feet, poop in the air, and fly right back inside. Our car is parked due south of two hives, and on winter elimination days, we find a scatter pattern of little orange dots on our windshield. Aside from that, everyone is slumbering in the cluster. Opening the hive will cause a drastic heat loss—too much for the sleepy bees to easily restore—and it disturbs them from their deep rest.

I know many new beekeepers whose bees have died over winter. When I ask what happened, they often recount a scenario like this: "It was sort of warmish, so I opened the hive to have a look. All the bees were dead on the comb, so I cleaned out the dead bees and took the remaining honey. I'll try again next year."

Let me take this story apart and tell you what most likely really happened.

> *"It was sort of warmish"*—Below 65 degrees is sort of coldish to bees. If you're not wearing shorts and a T-shirt, it's a cold day in bee land. Under 50 degrees, and bees no longer have the capacity to move.

"so I opened the hive to have a look."—Opening the hive releases any little pocket of heat they have. In winter it's really hard for them to create and maintain heat. When they need to make heat, which they do by shivering, they eat more honey, which depletes their saved stores. Over the course of the winter, less activity is better because it conserves their honey. Opening the hive in cold weather also breaks the propolis seal. This subjects them to drafts, and the bees aren't in the right mood to rally and repair it.

"All the bees were dead on the comb."—They may look dead, perched on the comb, hanging from one leg, soundless, unmoving, but if they've taken themselves down to torpor, they're just asleep. Again, torpor is a temporary suspensory state, not full hibernation. It is a metabolism slowdown that lets them go into a stuporous sleep that is hard to rouse from. Sometimes, even after the weather warms up, it can take the torpid hive three days to come back to full movement and function. If there is a frigid spell, torpid bees may look like they've frozen to death on the comb. When you look inside, you can't tell who is dead and who is asleep. The bees are deep in winter meditation, communicating with the sleeping flower spirits, and we don't want to disturb that.

"So I cleaned out the dead bees and took the remaining honey."—Oh no! At this point, you can't fix it. A few years ago I got a call from someone who had done just this. He brought the dead hive into the basement, took the honeycombs out, and put the honey into jars for the family. The next day he went downstairs and was surprised to find the combs crawling with bees. They'd been in torpor, and now that they'd warmed up and were awake, he was stuck. He couldn't put them back in the hive to cluster because he'd taken their honey, along with most of the comb, depriving them of food and warmth. He left nothing behind for the bees. They had no home to go back into.

"I'll try again next year."—If you do, please don't open the hive when the bees are at their most fragile. It is best to leave them without disturbance and let them wake up on their own schedule. Give them every good chance to make it through spring. If you want to know how much honey they have left, try lifting the back of the hive just an inch to see how heavy it feels. If you do this in fall when the hive is full of honey, and then do the same every few weeks until spring, you'll have a good idea how much honey is in the hive through winter and spring. (Lift it only an inch. Cold heavy comb is easily broken, so you don't want to tip it too high.)

Sometimes new beekeepers get nervous when they see a clump of dead bees on the ground outside the hive, but this is actually a good sign. It means the active house bees inside are exhibiting good hygiene and keeping the hive clean. The number of bees populating a winter hive constantly changes as older bees die off.

Every week I look at the hives on my deck and count how many dead bees have been carried outside the entrance. On a warm day a maiden will pick up the dead body and fly it a few feet or farther, where it won't attract interest from bee-eaters. Then she will drop it and return to the hive. Because this colony is on the wooden deck rather than in the grass, I can easily see the body count. Finding a hundred dead bees outside the hive on a warm day is typical. If a hive goes into winter with thirty thousand bees and comes out the other end with twelve thousand bees, the hive did okay. Even if eighteen thousand bees died over the winter, the hive is still viable.

No matter how much you want to look, control the urge! If the bees are dead in December, they will still be dead in March. On this topic, renowned beekeeper Michael Bush often says, "Good news is worth waiting for. Bad news will keep."

IN OUR OWN WORDS

Winter is the time of the cluster, dreamtime. The stores are in place, the queen's laying has quieted, and we draw together in contentment. The hive is sealed, suspended in a dream of hive images held by the

Unity. Like the swarm, it is a time when we dream of the Unity, holy days when we say, "I embrace and am embraced by the hive." Honey is passed and shared, and the memory of the sun's light is in our awareness.

We visit in our vision the place where we conceptualize the world, where we see a bee visiting a flower, collecting nectar into her. She takes the nectar inside, where she mingles it with her own life juices. The nectar she collects contains individual information about that flower and the collective information from that flower's family—its history, its intention for the future, and the knowledge of the land from which it comes.

She takes this nectar inside her and joins it with her own individual history, the memory and knowledge of this flight, as well as the history of her clan, which is contained in her enzymatic measure of her clan and her own helix. The little drop of nectar is a map of that flower and this bee's present being, which includes a sense map of the past and a directive for the future.

In winter we revisit the flowers through the taste and scents of the honeys. Each flower group has a signature we easily know. Though the flowers have died back and are now out of season, their flower essence continues within the hive and within us.

The honeys are opened in a certain sequence. Denser honeys are better for intense cold spells, and lighter honeys are for warmer weather. The nectars are collected and stored in a sequence so they can be eaten at the appropriate times.

Spring honeys fuel our building time, when we are most in the Song of Increase. Honeys gathered in late summer and fall are best for winter, when we slow down our activity, when we enter the dreamtime of cohesion.

Increase means expansion and creation. Cohesion means connection and community. Likewise these honeys have, for humans, medicines within their taste. Spring honeys fuel industry and activity that expand out into the world. Fall honeys draw us together in appreciation of community.

In winter dreamtime, we unite and communicate with the nascent beings who dream the next cycle's crops into life. When the hive is satiated, content, and in the torpor of winter's rest, we revisit the forage lines we laid in place during the harvest. These lines still exist in our collective memory, and ingesting these honeys replays for us the nectar paths. The winter bees may not have been privy to the nectar collection of each flower family, yet the knowledge is passed within the collective, like gathering around to read a map and sing together.

Moreover, the presence of each flower group's spiritual beings is acknowledged at this time. The flower spirits are in a torpor of their own in winter. The seeds are in waiting. The hive, however, is alive with the presence of these spirits. Dreamtime is a celebration of connection and appreciation, a time of deep spiritual union when we are intensely aware of the flower spirits. In this time the Overlighting Beings of the flowers are held in a cupped hand of gratitude as we send our appreciation—enriching and nourishing the flower beings and encouraging a fruitful next season. Even with biennials, the nourishment progresses the plant.

During winter's crystallization we see a multifaceted gathering of forces within the core of the light workers, our earth's helpers. All is brought together in unity, like the myriad windows in our eyes reflected in the multitude.

Plants with nutrient and mineral requirements that match the qualities of our area thrive in the soil. These plants send forth flowers whose nectar is gathered during pollination. The flowers, by way of their nectar, convey the etheric signature to the bees. The bees are constantly aware, even in winter. Thus, we function as the memory keepers, all year, of the vibration of the flowers.

The hive carries and holds the chemical, mineral, and nutrient stamp of that flower's family. It is more than this, though. This nutrient analysis is also the language of the planet and a treasury of plant knowledge. We interact with nature's intelligence throughout the year by recognizing and acknowledging the plants and the presence of plant beings.

This ongoing confirmation is crucial to plant evolution. In our awareness of the plant's makeup, we confirm, impress, and approve the plant's signature. Our stamp of approval injects a joyous life force into the plants every time we interact with them. This brings a trembling delight to the plant beings and creates joy throughout nature.

V

The Song of Increase

*The Blessings of the Swarm
and the Ascension of a New Queen*

The Song of Increase signals the most delightful time in the life of the hive, a time when everything in the hive is blossoming with right action. The bees revel in fulfilling their directive to bear increase into the world.

Two annual events are celebratory keystones in the life of a bee colony. One is swarming, the process by which bees split off from one hive to expand their genetic line out into the larger world to create another colony. The second event is a new virgin queen hatching within the hive and ascending into the role of queen.

This chapter explores both of these important events, as well as how beekeepers can support or inadvertently curtail the Song of Increase.

The Swarm: Birth of a New Colony

One evening I had a feeling that one of our hives was full and needed more space. Twice in the prior days I had seen big groups of bees come out at midday for what I thought were new-bee orientation flights—short forays the younger bees make in front of and around the hive that prepare them for transitioning to full-time foraging. But something about these flights seemed different. In a typical orientation flight, the bees do a lot of mingling and swinging back and forth as they test their wings while hanging in the air. These groups of bees instead flew out, hung in the air, and faced the hive, without a hint of back-and-forth movement, until hundreds of them were outside. Ten minutes later, they all went back inside. I mentioned to Joseph that we ought to add another box to the hive and give them more room to expand.

Early the next morning, before the sun warmed the air, seemed a quiet, peaceful time to add a box. This hive was a vertical Warre hive, in which new empty boxes are added on the bottom, and, when full of honey, hive boxes are taken off the top later in the season. I placed an empty box on the ground next to the hive so I could put it underneath once Joseph had lifted the stacked boxes from the bottom board.

When Joseph picked up the stack, I expected to see the empty wood floor underneath. Instead, the entire floor of the hive was thickly covered with bees, all facing toward the hive door. The gathered bees looked like a cadre of suited executives, each with a briefcase in hand, waiting for a train in controlled anticipation. They were perfectly still when we lifted the hive body up. Not one bee flinched.

We were weeks too late to add the box. The hive had long since determined their space was full, and they stood at the ready, patiently awaiting the signal that would tell them to begin the swarm. Sure enough, when sun hit the hive in late morning, they poured out and departed in an enthusiastic swarm cloud.

I have noticed that swarms are very good predictors of sunny weather. When my phone rings with a few swarm calls before noon, I know that means there will be a break in the weather. I tell my husband it's a great day to make hay because the weather will be dry enough to leave it on the ground for a few days before baling.

I won't say that a swarm's weather knowledge is 100 percent reliable. I do find wet swarms occasionally, but that could happen because a swarmed hive didn't find a new home quickly enough and hung in a tree so long that the weather changed. Once in my own bee yard, I saw a hive swarm to a nearby tree and park on the tree limb for an hour, waiting for their new home to reveal itself. When the sky suddenly clouded up, they matter-of-factly moved back into the hive and waited inside, dry and safe, until the next clear day.

What prompts bees to swarm? Each spring, as the flowers come out, colonies go through a growth spurt called *building up*. Longer days and warmer weather draw the bees out into sunshine, where blooming trees, bushes, and field flowers beckon. By nature, bees are industrious. They gather pollen and nectar to feed everyone, and they build comb to fill with food and new bees. The queen lays thousands of eggs to increase the colony's population. The intent is to completely fill the hive, and most colonies do that well. Knowing the colony is nearing peak capacity and ready to create another entire hive thrills them, and activity ramps up even further.

Every day the house bees prepare the empty cells for the queen to lay eggs. In times of abundance, when plenty of flowers are in bloom, the colony may expand the honey area to wherever they find empty cells, including the brood chamber. When a colony has no more brood cells to fill because they are all full of honey, the hive is called *honey bound*. There simply are not enough empty cells for the queen to continue laying new brood. The colony's growth stops and can even begin to reduce, which is not what it wants in the time of increase.

When the bees know the hive is nearly full—when there is no more room to place more eggs, pollen, or honey—they begin preparations for swarming. These preparations are extremely exciting and leave the hive humming with activity. They make sure the departing queen has laid enough queen eggs to ensure that a new queen will hatch and replace her. Scout bees, meanwhile, will have started looking for a new home. When they find one, the swarm will depart the hive, land on a branch for a few short minutes to make sure the queen is with them, and then fly off to their new home and get started building comb.

A swarm is far more organized than one would think. The intention of the swarm at the start is to create chaos, and from the outside, that's just what it looks like. If you are lucky enough to stand inside a swarm, you will see bees flying in every direction in an ever-expanding sphere. Many times I've walked into swarms and been amazed at the complexity of their undertaking. Each bee flies in looping circles, keeping a near-miraculous even distance between and around each bee in the sphere. As I've moved inside the swarm, never has a single bee mistakenly flown against me. Each healthy bee knows the exact location of every other bee, tree branch, person, and any other object near the swarm, and accommodates itself around those objects. (Except for drones, who swarm-fly with a jaunty, bouncy joy, and I do see them bump!)

Bees are at their most gentle during a swarm. They have no territory to protect and only one immediate mission: to conceal the queen as she flies in the midst of them. The chaos of so many flying bees is meant to keep anyone from knowing which bee is the queen. You'd be hard-pressed to see her in the clouds of thousands. The bees create enough commotion that their precious cargo, the beautiful queen, stays hidden among the many, protected from anything that might cause her harm.

Once everyone is in the air, they raise the energy even more, doing the bee equivalent of whooping and hollering. The scene is pure exuberant happiness.

During swarming, a magnificent event occurs: the queen's fertility is renewed. When the queen flies with the swarm into the light of the sun, the sunlight replenishes her hormones and ensures her reproductive ability for the coming year. In this way, through swarming, the queen keeps her fertility intact for another year.

Once the queen is enswarmed and has flown in the sun's light for a while, the swarm's purpose changes. Now the swarm seeks to find a place to land and gather themselves together. The transition happens so quickly it is startling. One minute, thirty thousand bees fly about in what looks like disorganized bedlam, and then suddenly there is a focus point, such as a nearby tree branch. They fly to it and begin landing one atop the other. As they land, they form into a hanging, elliptically shaped cluster, like a soft football. The first layer of a few hundred bees grips the branch, and each successive bee grabs onto another bee's legs until they become one hanging mass of bees. When all the buzzing bees have landed, they quiet into a soothing hum, resting in place, waiting for the scouts to find and communicate where their new home will be. The swarm cluster won't move again until a new home is discovered.

This may take half an hour or, if no potential home seems right, a few days. During this time, they patiently wait, docile and meditative, in quiet rapture.

IN OUR OWN WORDS

Swarming is an expression of gratitude for the colony, a proclamation of work well done. The queen toils in darkness all year except for this one brief time when she emerges into the light.

An ascended queen stores light like a holy sacrament within her. She doesn't need a lot, but she does need to come into the light once each year to renew her fertility. Her brief annual flight reconnects her with the sun. The sun's light on her body stimulates her reproductive system—a symbolic remembering of her mating flight—and renews the fecundity, the life force, within her. This is not merely a random moment of brightness flashing upon her. The entire mature community participates and is integral to this renewal.

Before we leave our old home, we who are departing fill our bellies with honey, enough to last through our journey. We make joy, like a bon voyage, a celebration. The hive is filled with anticipation and exhilaration. Everyone who is leaving moves toward the open door in high excitement. We pour out of the hive like water.

When we leave the old hive, we come out and fly in rings and loops. We color all the spaces with our presence. This is an expression of joyful excitement at the imminent increase we are embarking upon. In the flurry and whirling, we create a veil for our queen.

And now the queen emerges into the swarm. Sometimes the queen has done this before and has a memory of flying and floating in the air, surrounded by the hive bees. The queen glows in the sun's light. She flies freely in the mass of bees, the entire hive surrounding and concealing her as she drinks in the sun's nourishment. In swarming, the queen mates with the sun again, a joyful orgasmic culmination and celebration of purpose, duty, and destiny. The sounds and movements are foreplay to the swarm's orgasm, each step fully expressing the hive's mission of continuing fertility. The swarm provides safety for her renewal. The queen opens herself energetically and physically, inviting the coming year's fertility. The sun's light reaches into the queen, initiating a chemical process that unlocks and vitalizes the coming year's generation of sperm and seed, and thus renews the hive. In her beauty she flies at the heart of us, just as she dwells within the heart of the hive.

This is a celebration of our increase, and nectar-laden we each are, the sweetness of life within our bellies. We especially like the spring honeys for our journey because they center us on a map of our lands, immersing us in the scent and flavor of the plant life we serve.

As we fly out, each bee has an energetic cushion around her, and we are intensely aware of all the bees around us. During a swarm, we have heightened perception. Even though the swarm is moving in every direction, each bee is aware of all the cues going through the swarm. In an instant, we sense that everyone who is coming is with us. In that moment, flying in all directions becomes flying in one direction.

In the landscape a spot is picked to alight, and we fly to that spot and coalesce into the swarm body. We land and gather with no protection but our number. Clasping each other, the central bees hold the branch. Layers of bees land and take hold of the bees beneath them. Each and one, we are breathy, cheerful, adventurously elated.

Upon alighting on our branch, we embrace. We clasp each other, creating an interlinked mesh, hand to foot, layering as a gilded swarm. We sing, "All is well. We are in the hands of God."

Communing with Swarms

There's a telltale sound a swarm makes, and after all these years, my ears are tuned to it. More than a few times, I've been outside and far from the hives when I suddenly heard a voluminous, hissy thrum calling to me. Whatever I was doing falls off my to-do list, and answering the swarm becomes my higher purpose.

If the swarm has made itself known, a beekeeper has two choices: let it go or collect it. If you let a swarm go, the bees will find a hole in a tree way up high or some other inaccessible place. Collecting the swarm allows you to move them into an empty hive that will, if they like it and decide to stay, become their new home.

Gathering a swarm and inviting them to live in one of my hives is one of the most sense-enhancing, gleeful tasks I know. Often I find them on a tree branch, and moving them is easy. I place a box directly underneath the swarm and give the branch a good shake. The majority of the bees fall in a clump into the box. A few hundred may rise up in a brief buzzy cloud, but if the queen is already in the box, all the bees outside the box will find their way inside to be close to her.

Every so often the swarm lands on something that's not as easy to shake as a tree branch, and I have to get creative. I have removed bees from a cyclone fence, inside a pipe, under barn wall shingles, in a porch sconce, and inside the bottom compartment of a barbecue grill. A friend of mine once collected a swarm from the backseat of a derelict Volkswagen Beetle.

Most beekeepers want to grab every swarm they can—free bees!—and in my early days, I collected plenty for myself and my friends. Those first years of swarm collecting made for exhilarating bee seasons where I got very little other work done. I was swarm-swooned.

These days I have a good amount of bees on our farm, and I don't feel the need to gather up every swarm that crosses my path. I've intentionally let many swarms go. Not all swarms want to be caught. This past spring I was fifteen feet up a tree for three hours with an empty box in my hands. Instead of catching this enormous colony, I sat transfixed with their beauty until they finally flew away. As they flew off toward the tall cedars, I blessed and thanked them for spending time with me.

Sometimes, though, I have a particular fondness for a certain colony, and instead of waving goodbye when they swarm, I collect them and give them a new house to inhabit. If they came from my bee yard, I place them a hundred feet or more away from their old home so they're not living right next door to their family.

"How does one collect a swarm?" you might ask. If the swarming bees were on flat surfaces, I used to gently sweep them with a soft brush. But early on I realized bees don't like being rushed or pushed, so now I scoop or direct them with a long, sturdy turkey feather. Because we have a farm, we always have plenty of different-sized feathers, and bees seem to like feathers just fine. I say, "Here comes the bus. If you want a ride over to the other bees, jump on," and to my delight, they climb aboard.

Once my bees swarmed onto a very leafy wisteria in our front yard. I couldn't figure out how to get them off the vines and into the box. With gloves I wouldn't be sensitive enough to know if I squeezed or injured a bee. I had handled over a hundred swarms at that point and knew how calm they could be, so I decided to work bare handed.

I tenderly eased my hands ever so slowly into the swarm, and I instantly felt myself become completely present with them—not a thought in my head but the mystery of bees. I felt the gentle movement and benevolent breath of the cluster as they sweetly opened a narrow path for my hands to enter their body—a holy and gratifying moment.

Moving a swarm with bare hands isn't something to be done lightly, or it likely won't turn out well. Be thoroughly familiar with swarms and capable of maintaining a sustained meditative state before you do this. They are small innocents and require that we care for them with likewise virtue.

Swarm Scouts: Seeking a New Home

Even my best laid plans for moving a swarm don't always coincide with what the bees have in mind. In my bee yard last week, I collected a good-sized, healthy swarm. I got the bees all nicely housed in a top-bar hive and stepped back to have a look. I know they prefer that their new home be a good distance from their old home, but they were just settling in. I decided to leave them there until their scouts returned, so then I could move all the bees together to a new location further away. About twenty minutes later, the swarm suddenly mobilized a quick departure and abruptly left. Off the bees went in a buzzy, seventy-foot-high cloud. My farm interns and I followed the swarm, but lost sight of them when they went through a forest of tall pines. Though running full speed, we simply couldn't keep up.

The scouts had found a place that was more to their liking. The speed of their leaving tells me they had probably decided upon their new place even before they swarmed. My swarm collection was merely an inconvenient delay.

IN OUR OWN WORDS

Once we have landed, our scouts begin seeking a new home. If some scouts have already found places in the area, they direct the other scouts with their dance, so the other scouts can check it out and come back. Even though a scout bee goes out and finds a potential new home and reports back to the swarm with its location, that bee has no sense of individuality with that task.

Before the scout bees leave the swarm body, there is an energetic membrane around the swarm. When they fly off to scout, they don't disengage from that membrane; they extend it out to the locations they are scouting. The swarm experiences the location directly on the scouting trips. The membrane is like a golden bubble.

The scouts are our senses as they search for our home. They enter each possible home and communicate back to the swarm information of each place's suitability. When a scout enters a potential hive home, she stands within the cavern and emanates a projection of this place filled with comb and in its fullest expression. She is not looking at empty space; she is seeing the place as a full working hive. She notes especially the movement of air within, how readily this can be modified to protect the brood nest. Though size is significant, the air within is even more so, as the brood is our utmost concern. So we also want a hidden and defensible entrance.

The sound of the vibration inside—the resonance—is important because communication is so vital to us. The warmth of the space is important, as is there being enough room to expand the colony as it grows. In the wild, most hives are small. We seek to fill the hollow of the home we have chosen. We often choose a smaller home over a large one because we know, regardless of the year's bounty or dearth, we are more likely to succeed with the frugality of our choice.

A large home in a dearth year is a danger to us. We would struggle to keep the warmth throughout without our stores of warm honey surrounding us. A year of bounty means we are full to every nook, and our survival is assured. Though it seems that our colony's size is determined by the volume of space around us, space is not our first desire. We choose a home taking into consideration average years and dearth years. Though a large home is a pleasure when the year is a bounty, we do not seek the home that succeeds only in bounty years.

You who keep bees measure success by what we produce, scoring by volume and size, but this is not our measure. Who gives more, or most, is an unbalanced system. If the colony cannot sustain itself from year to year, with every aspect in harmony with the next, that colony is at risk. Our tasks are all of equal importance, and any lack in competence brings about failure.

Within our home we also prefer a fall-away so that anything that should not be in the hive or upon us can fall away and keep us clean and healthy. When we live in the tree, what we groom off falls down into dried comb and litter. Within the hive walls, we build an envelope of our medicine, the scented propolis, on top and around us. We seal all but the bottom, where the fall-away serves our needs through the community of life that processes what we no longer need. The fall-away is part of our medicine and a key to how we maintain the hive's health.

Our hive, our home, teems with life. These other life forms inhabiting the fall-away have their own purposes and prevent an accumulation of detritus beneath us. These tiny members of our biome are co-participants in a very precise community that evolves with each hive home's specific needs. Though we have some aspects in common, the evolving work of the fall-away inhabitants brings forth an exquisite balance of transitional directives. Nature marries us to those with whom we share our home: dismantlers and movers, creators and orchestrators. Each works within this fine balance so that all emerges in its own time.

Knowing this, you who seek to provide us with homes may want to reconsider what makes a thriving hive—that is, our competence in all tasks and in the evolving community in which we exist. The measure then is not our product but the process of our harmonious living and our respectful relationship with those around us.

As each scout returns, she dances the information and the other scouts take off to have a look. The visit is, for each, a time to envision the working hive, and news of the suitability moves through the swarm. We all have an accumulating image-sense of these possible homes in the projection details as they are inspected and communicated.

The dance gives detailed direction to where the potential home site is, but the projection has already been delivered. Within the swarm, each of us perceives, from the perspective of the "wholeness of being the hive," our likelihood of thriving within this new home. We see ourselves already living there and the ease and comfort we will share within that place.

As news of these locations passes through the swarm, we find one location more desirable. The communication of each location has been shared, so we are all part of the selection. There is no separation from the knowledge, as one bee entering that place is, by nature, the senses of the rest of us. Her emanation within the space is carried to us all. Thus, there is no need for the swarm to visit each possible location. The scout and other scouts who accompany with follow-up visits express in their emanation the suitability of each location. The emanation expressed within that location is the first layer of how we will create our home within this space.

When a location is less suitable, we see that, too. When follow-up visits continue to express the shortcomings of a location, we lose interest and that location recedes in our consciousness.

When enough scout bees extend the membrane to a specific location with a sense of approval, the knowledge of our new home emerges in our awareness, and we begin our journey to the hive-to-be. In the air of this flight, we and our queen lay a scent trail for scouts who will later follow.

Once inside our new location, each bee adapts to a new role, and we begin our work. As mature bees, the swarm bees have outgrown the ability to make wax like younger bees do. Yet because the new hive needs wax makers, we are capable of turning back time in our development and once again produce wax. When our hive has a need, we come up with a solution.

Separated from the Swarm

My experience is that scout bees, upon returning to the swarm site, can miraculously follow the scent trail left behind after a colony departs. I have seen scouts from a swarm land where their hive mates had rested earlier in the day, take a few steps around the site, and then rise into the air and follow a scent trail to catch up with their colony. If the day has been calm, the whole hive can be reunited.

One day the weather had been particularly windy, and the scent of the swarm trail was too dispersed. At dusk I found a few hundred scouts clustered on a forked tree branch where the swarm had been, and I collected them in a hive box. I considered taking a spare queen cell from another hive and giving it to them, but there really were too few bees for them to form a hive of their own, and by then I'd given away the swarm they belonged to. I put a few bars of honeycomb in with them, and the bees happily stepped onto the familiar comb surface to wait with hope for their colony's return.

A queenless cluster of bees has a certain sadness to it. For a few days the little band of scouts foraged and filled the comb with nectar, trying to keep up their bee behavior. But as days passed, they appeared dejected, and each time I looked, I felt a big sigh come from them— completely appropriate behavior for a queenless bunch. I wondered what to do with this forlorn little group of scouts. I thought about another hive in the bee yard that could use a few more helpers and decided to add the scouts to it.

Ordinarily, I would merge two hives by placing a few sheets of newspaper between the hive boxes to get each group used to each other's scent, and let them merge on their own terms. But these bees looked so downhearted I didn't imagine they would be a threat to the other hive; they'd been apart from their colony long enough that they didn't smell like another hive's queen anymore. My intuition said that instead of letting them have a gradual introduction, I should try something different.

I inched the hive cover back, exposing two bars. That way I didn't alarm the established hive, and I could easily see if what I was about to do would work. I collected a few scout bees with a clear shot glass and a small index card—my best single-bee-catching method—and released each contingent of bees into the hive by ferrying them with a feather.

When these first few abandoned bees made contact with the home bees, there was a lot of touching and checking them out. Then the home bees walked back to their tasks. This little band of bees did not hesitate for a moment upon entering the new hive, and nobody in the hive questioned their arrival. As I put the feather or the glass next to the opening, each bee dashed through like it was their long-lost home, and the little crew of bees slipped into the hive. I felt sure I could see them grinning, like sailors lost at sea who'd suddenly found land. Such good fortune! The new bees immediately got to work, integrating into the new hive in just a few minutes.

This move wasn't a big deal, but moments like this are the ones I find so rewarding. A few hundred bees would survive, and I had something to do with it. There are so many times when I wonder if what I'm doing is the right thing, and here, watching the lost bees sprint inside, I knew it was.

Drumming the Bees

When Gunther Hauk, a fellow biodynamic beekeeper and author of *Toward Saving the Honeybee,* was at our farm teaching a biodynamic beekeeping class, he briefly mentioned that it is possible to use a drum to call the bees into the hive. I searched for "drumming bees" on the Internet, but most of what I found was about drumming on the hive to get bees *out,* an action called *tanging,* which I did not want to do.

One day a swarm from our farm landed forty feet up a cedar tree on an outer branch. I wanted to move them into an empty wooden hive I'd set up eighty feet away from the tree. But I didn't have a pole-with-a-bucket long enough to reach them from below, and even if I were to lean a ladder against the tree trunk, they were too far out on the branch for me to reach. I'd never seen someone drum to call bees into a hive, but with nothing to lose, I decided to give it a try.

From what Gunther had described, the process is simple: I just needed to stand next to the empty hive and use one of the wooden bars to hit the side of the hive in a steady beat. I decided to drum in a four-part rhythm—hard-soft-soft-soft, *bap*-bap-bap-bap—like a child's tom-tom rhythm. It took me a while to fall into the rhythm. I kept saying to myself, "This is silly. The noise will drive them away. How could this possibly work? I'm lousy at keeping the beat." At four minutes, I was bored. After six minutes, I was really bored. But I continued.

Fifteen minutes passed—plenty of time for me to wonder why drumming might work. I couldn't come up with any levelheaded reason why bees would come toward the drumming sound, but I pounded on. I mulled over the observation that I was not metrically competent enough to keep even a simple beat, but I let the thought go and continued *bap*-bap-bap-bapping.

Then, in one startling moment, the swarm became a dense bee cloud, lifted up out of the tree, and began flying straight toward me.

Fifteen feet above the ground, twenty feet from me, they suddenly spun around and turned back. Most likely the queen wasn't with them, so they all went back to the tree to re-gather around her. But I was excited. I'd drummed, and the bees noticed!

How did that happen? Why would drumming be important to bees?

After I calmed down, I asked the bees what they look for in a new home. They said they seek resonance: "Our communication within the hive is important. We need good sound and vibration inside."

Wild bees prefer to live in hollow tree trunks. In treeless landscapes, clefts in rocks will serve, and cavities in human habitation may also be deemed suitable. Scouts would need to visit every single tree to find an opening to a hollow inside; otherwise, how would bees know a tree is available? I kept thinking of their answer—resonance—and wondering how nature would convey that information to a colony. It occurred to me that a branch banging on the tree trunk as it blows in the wind or a woodpecker rat-a-tat-tatting on the tree looking for insects would readily telegraph to the bees, "There is an empty chamber in this tree you might want to come look at."

Later that day I borrowed a telescoping pole from my neighbor and managed to get the swarm down out of the tree and into a bucket, but the idea that drumming might work had latched onto me. Later in the week, when I brought home another swarm, I again gave drumming a try.

Bees can easily get tangled in tall grass, so I laid a white cotton tablecloth on the ground. Using a wooden shingle, I made a ramp from the cloth up to the hive entrance. Then I upended the box, dumping the bees onto the cloth, and I commenced tapping on the hive. The bees milled about in every direction. I drummed the familiar *bap*-bap-bap-bap rhythm. After a minute or two—totally surprising me once again—all the bees turned their heads toward the entrance and started walking up the ramp and inside. It took fifteen minutes for twenty-five thousand bees to climb the ramp and get inside, at which point I carried the full hive up to its permanent location.

I must admit, drumming was much more fun than dumping a boxful of bees into the hive, as I'd done in the past. Because it seemed to work so well, I now sometimes use drumming as a way of inviting a swarm to choose—on their own—to move into the hive I am offering them.

Recently in my bee yard, a swarm found its way onto the edge of a picnic table. I set an empty hive on top of the table and drummed them up and into the hive. It was quite remarkable. Maybe they would have found their way inside without the drumming, but I like to think the rhythmic vibrational sound was the perfect invitation to check out the hive and move in.

The Ascension of a Queen: Blessed by the Light

After most of a colony departs the hive in a swarm, taking the colony's queen with them, the bees who are left behind need a new queen.

As already discussed in chapter II, a week before a swarm leaves to make a new hive, the departing queen deposits eggs into vertical queen cells. A week after the swarm departs, the group of twelve to fifteen virgin queens hatch within a few days of each other. If only one of these hatchlings can become queen, why are there so many? Nature encourages redundancy so there is less chance of failure. Having a dozen backup queens ready in case the first one (or several) doesn't survive is wise.

The young virgin queens emerge from their cells in the darkness of the nursery. They crawl along the walls, exploring the comb and touching the other bees as they work. A week later, the virgin queens are strong enough for their marital flight, so they find their way to the hive entrance, to the light. At last they stand on the front doorstep, basking in the warmth of the sun.

A virgin queen is treated just like any other bee and gets no special treatment from anyone in the hive until she has mated. If she asks, feeder bees will share food with her as they do with other bees in the colony. That first week after she has hatched is spent in waiting. Her body needs time to mature before she is ready to go on her marriage flight.

A virgin queen leaves the hive with a small contingent of maidens who are familiar with local foraging areas and can help her find her way back home after mating. She searches the sky for the area where the drones gather each warm day, waiting for virgin queens to arrive. These drone congregation areas bring together drones from different hives, thus offering mating queens an expansive range of genetic diversity. Drones from as far as two or three miles away gather in these common regions, which the bees call a *scarp*. On sunny days the scarp is fully populated with drones from many colonies. The eager drones arrive an hour before the virgin queens appear. The drones fly in long looping circles, their enormous eyes wide to the skies, watching for a virgin queen's arrival, reading the air for the seductive alluring pheromone that announces her readiness to breed.

Once the big-bodied drones see a virgin queen, they chase after her. Only the strongest, most determined drones will catch and mate with her. The first drone who catches her grabs her with his legs and inserts his phallus into her, at which point she contracts her abdominal muscles, drawing the sperm mass into her, and pops the joint between his phallus and his abdomen. The drone falls to earth, his reproductive organ still implanted in her. The left-behind organ pulses with an ultraviolet light readily seen by other drones, who hurry to catch and mate with her. One drone after another, a dozen or more, each with his own unique genetics, mates with the queen. As each drone donates his sperm, the new queen gathers an enormous range of potential behaviors and characteristics for her offspring. For the rest of her days, the sperm from this glorious mating frolic keeps her fertile and fruitful. Once her marriage flight is completed, she will never again mate nor race headlong into the sun.

When a mated queen returns to her hive after this flight, she cleans herself up and then seeks out and attempts to kill all the other virgin or mated queens she finds. Once the queens are reduced to one, the remaining queen is welcomed as the hive's new reigning monarch and given a full royal court of attendants to serve her every need.

Occasionally, this last scenario works a bit differently. If the hive truly is overflowing with thousands of extra bees, the Unity may imagine the colony capable of throwing off a second and a third swarm to create even more new colonies. In that situation, a contingent of guardian maidens may form a wall around some remaining unhatched queens to prevent the new queen from killing them all, reserving a few who will soon also become swarm leaders.

Once she has ascended the throne, the new queen begins her role as mother of the colony. Except during cold winter months, she will lay fifteen hundred to two thousand eggs a day. She continues creating new life for five to seven years. Barring accident or illness, she will remain the reigning queen until she and her grown colony swarm the following year, leaving behind new virgin queens to take her place.

This is what is conventionally understood about a queen's ascension, but there is so much more to this story. From the bees' perspective, it is an epic journey of sunlight, longing, becoming, and the mingling of earth consciousness and body consciousness.

IN OUR OWN WORDS

When the virgin queen emerges from her cell, she moves first downward to the earth. Unmated and newly born, she is nearly invisible to the hive, her role still empty. She moves within the hive with curiosity, but not yet commitment. As virgins, multiple queens may pass through the hive corridors, but, until mated, none will ascend as queen. Unmated virgin queens are singularities within the hive, not yet manifest, nebulae in the hive's periphery. The untouched and virginal queen is a promise as yet unsealed.

The young queen wanders the halls, free of the weight of her future duties. She will rest for a few days, gaining the strength needed for her marriage flight.

As the virgins walk the hallways, waiting for the right weather, they may emit a sound, a quick burst of energy that speaks to the sun. It is a call to the sun, a promise that she is coming. The sharp sound made by the virgin queens opens the clouds. This vibration of focused diffusion turns the heavy mist to rain, or if the mist is spare, it opens the light. The noise the virgin queens make is called a *harken*—a solar harken, an announcement.

This spike of energy aligns molecules to open the air, causing clouds to burn off. The virgin queens call to the sun, over and over, sending the sharp sound out to the gray skies. The vibration goes out from the hive and cuts a brief horizontal slice in the clouds. This pushes moisture-laden clouds or fog into more condensed areas that, once made heavier, may break the weather. The virgin queens call to the sun, enlivening areas of stagnation in the air, bringing movement to the atmosphere. This relatively rare sound calls to the water sylphs to help free them from their dream time and open the door that brings them into the world.

When the weather is right, she finds her way down to the hive entrance, to the light. At last, she stands on the front doorstep, basking in the warmth of the sun. The virgin queen's flight is a marriage flight, the only time she mates with the drones.

The earth has a light of its own, deep within it. Across the land there are energetic openings in the earth's surface called *lumens,* where the earth's light and sunlight joyously communicate with each other, singing earth's planetary signature and contributing to the blended symphonic sound of all the planets' chords.

Inside the sparkling cone-shaped lumen, the earth focuses on the sun's light in acknowledgment of the gravitational relationship between the earth and sun. The natural mating flights of honeybees take place in the lumen, in portals visible to drones and queens.

These light-emitting areas where the sun speaks with the earth are gravitationally uplifting places, and the drones know this. The drones' wondrous eyes are capable of seeing the lumen glow—another way their eyes differ from the maiden bees. Within the lumen, the gnostic drones spin and whirl, taking great delight in the lightness of the air that buoys up their heavy bodies.

Drones are drawn to a specific place within the lumen, often a few hundred feet up, where a higher focus of intention occurs. Humans named these locations drone congregation areas, but to bees, the drone site is called a scarp and is known to all the colonies in each area. Drones from all the nearby hives spend much of their afternoons drifting in the scarps.

When the sun moves to midday, the drones emerge from their hives and hasten to the scarp. The drones fly an embosoming blanket, drawing circles and figure eights over and over, laying the sheet of the virgin bower. They weave a bed of prayers, a holy sanctuary of tremendous knowledge and healing.

When the virgin queen emerges from her hive, she surveys the landscape looking for the lumen. When she sights it, she spreads her wings and flies directly to it, entering the lumen low in the light cone. Once she is in the lumen, the lumen's glow turns on everything in her, illuminates her, even changes her color. The light opens and gives her knowledge of her role.

In her hive birth, the queen was born into her body. The lumen birth now bears her into her life purpose. Upon entering the light, the queen is quickened. Blessed by the light as in a christening, she is made holy.

At the base of the lumen, she aligns herself over the opening and focuses her intent on the ascent. The beacon of light that emanates from these earth energy points is charged by her stimulatory energy as she flies up and into the lumen.

Though humans may think the purpose of the young queen's marital flight is only to mate, the new queen flies with a more developed intention. Her maiden flight has deeper purposes that raise her into the sky.

She was born with a fundamental imperative to deliver her hive's message to the sun. This message describes her incarnation and the conditions about her as she rises to the mating. In delivering this message, she unlocks and begins her métier as mother of the hive. She shares this communication and her hive's physical and spiritual genealogy as she opens herself to the sun.

Her next message is to the heavens: each virgin queen carries a distinct sound, an audible vibration of joyous expectation. When the queen pierces the earth cover during her mating flight, she shoots a bolt of knowledge—all that the hive is aware of—in a message from earth, launching it outward to the planets, conveying the progress being made on earth.

In our creation song, the drones sing of the queen's lumen birth to the pips, so the young bees hear this part of their history. The drones sing about the lumen, how the queen is blessed and made holy as she ascends in the lumen's earth light, and the knowledge of her purpose comes into her.

The lumen causes her pheromone to bloom, expanding the scent of her holy purpose. Beginning below the blanket of drones, she rises at great speed until she pierces their circling field. She flies straight up toward the sun through massed layers of drones, spreading her scent. When the scent enters a drone, he becomes singularly focused on catching and mating with the young queen.

She flies upward toward the sun as the knowledge unseals and opens in her. Information pours in. She glows and becomes stronger. The drones, with their wonderful eyes, see the light filling her. The drones see things maiden bees are not privy to, and this is one—witnessing the light filling the young queen.

As the drone launches upward, the gravitational force pushes on his abdomen, forcing his phallus, till now unbidden, down and out from the base of his abdomen. Ready for mating, the fastest drone reaches the queen, joins her, delivers his seed in the tumult of creation, and falls backward to earth, dying before he reaches the ground. The next drone reaches her, removes the first drone's remaining part and joins her, each drone successively entering her and delivering his seed.

This union of each is a spark, a cumulative bright light that bursts into the atmosphere. Each consecutive union renews the message to the spiritual realm. The sky glows in the lumen, the holy point where heaven and earth merge in creation.

The queen's flight and the drone's mating build the strength of this outward beam. The queen exerts her energy toward the sun, and if she continued onward, she would exhaust herself in flight. The drone reaches and embraces her. He fills her with his seed. She flies on toward the sun, becoming more whole, more holy, until at last, the weight of her knowledge and the compass of her seed turn her earthward, toward her hive and her new life as queen.

The scent of the mated queen tells the hive she has, at last, arrived. As she enters the hive, the last drone's appendage is removed. She sets herself in order and dispatches any remaining unhatched maiden queens. If necessary, she challenges and defeats any other queens who also had a successful mating day. Once the details of her preparations are handled, she is ready for the hive to acknowledge her ascension, which they do. The hive immediately and totally aligns itself to her role and her purposeful presence. A calm enters the colony, and peace fills the land.

These mating sites—the scarps—are earth acupuncture points, each a fountain of renewal for the earth, sending knowledge out to the heavens and coursing an acknowledging energy deep into the earth. This piercing energy is a union of heaven and earth. It is of great import that these earth acupuncture points be stimulated by the natural union of the bees each year at their right time.

The lumens are tremendously holy and often near shrines, as they are themselves holy entry points. People who move through these places feel uplifted in spirit and grounded in purpose.

Conventional Beekeeping and the Song of Increase

Swarming and the ascension of a new queen are peak events for a colony, and the bees revel in them. Yet both of these events are often denied them in conventional beekeeping.

How Thwarting Swarms Hurts Colonies

One reason swarming is prevented is that it constitutes a big decrease in a hive's workforce, which means less honey and, therefore, less money for beekeepers. Another reason for denying the colony the joy of swarming is supposedly to protect the public, who is often misinformed about swarming by inflammatory media stories that reinforce fear of bees.

But not being allowed to swarm creates problems for a colony. Swarming is nature's way of breaking pest and disease cycles. When an infested swarm moves into a new home free of infected comb, there aren't any cells ready for laying pips yet; therefore, there is no place ready for mites to lay their eggs. The bees start building new comb, but it will take a while for that brood chamber to be built and made ready for the queen to start laying eggs—long enough for any old mites who traveled with the bees to die off. That throws the mites' life cycle out of synch with the brood's hatch cycle, effectively putting the mites out of business and giving the bees a fresh start. When conventional beekeeping thwarts swarming, it prevents bees from ridding themselves of mites and other pests.

Also, as mentioned earlier in this chapter, a queen who is prevented from leaving a hive with a swarm often loses her fertility in subsequent years. During swarming, the queen's flight in the sunlight, in the presence of the sun, is what turns on her hormones and renews her fertility.

If she can't be in the sun with the swarm, her hormones don't cycle, and she becomes infertile.

The Problem with not Allowing Colonies to Develop Their Own Queens

In conventional beekeeping, young queens are not allowed to fly out and mate with wild, strong local drones. Instead most queens are artificially produced to become new queens through a process that allows beekeepers to create hundreds of queens at a time, far more than a colony would on its own. Often these queens are artificially inseminated, which lets queen producers control genetics, keep the drone bloodlines pure, and mass-produce new queens.

Gunther Hauk is an advocate for natural queens. His learned book, *Toward Saving the Honeybee,* explains how natural queens significantly differ from the industry-standard, artificially produced, or grafted queens that most conventional beekeepers depend upon. According to Hauk, "The single most serious factor causing the lowered state of health and vitality of the honeybee [is] artificial queen production."

Let's look at how bees naturally create a new queen when they need one in a hurry, when their old queen unexpectedly dies. When the queen dies, no one is left to lay maiden eggs. Without new babies being born, the entire hive will quickly die. The bees need to act fast, while there are still one- to three-day-old, and even up to five-day-old, maiden eggs incubating. The bees quickly find a young egg and change her six-sided horizontal maiden cell into a larger, round, vertical queen cell. From then on, they feed her royal jelly and withhold the pollen that maidens eat. The pip's new cell and royal jelly diet, miraculously, turn the maiden egg into an *emergency queen.*

Despite the pip hatching as a fully functioning queen, the colony knows that at the moment her egg was first laid, that queen was supposed to have been a maiden. She was altered in her cell to become a quick replacement for the lost queen and was not a true royal-from-the-first-moment queen destined for queenhood.

Emergency queens may have vigor and may build brood quickly at the start, but they are not known for longevity. Emergency queens commonly lose their fertility within three to twelve months, sometimes sooner. If the colony is left to choose, they usually replace an emergency queen within her first year by a process called *supersedure,* in which the bees will raise a true queen to take her place. The supersedure process is calm and gentle, with no disruption to the hive. The new "true" queen moves smoothly into her work, without violence, as the emergency queen goes into retirement. I've opened a hive and found an old, unattended emergency queen in the upper chamber, strolling about without worry.

Nearly all conventionally raised queens are actually emergency queens, and their fertility is unreliable. Loss of fertility doesn't happen nearly as often with naturally raised queens, who mate with hardy drones on their marital flights. Natural queens can continue laying eggs for five years, and some say up to seven years.

Alas, many hives are forced to operate according to a beekeeper's agenda. Sadly, that agenda often prioritizes more honey production and pollination services instead of putting the evolutionary needs of the bees first.

VI

The Song of Abundance

The Generosity of Bees

When people learn that I keep bees, they always ask if I collect lots of honey. Honey is a superb and generous gift of the hive, and I do take some, but not much.

Beeswax is made by the young virgin bees, a detail early church leaders took great delight in. I keep a beeswax candle lit when I work in my studio as a way to acknowledge and honor the enlightened labor of the bees.

Propolis is medicine for bees, and it has a respected history as a medicine for humans as well. Antonio Stradivari used propolis to make varnish for his fabled violins.

Though most people try to avoid a bee sting, the beneficial venom can also heal inflammation and illness.

And then there are the bees themselves and the profound teachings they offer those willing to be with them in kind observation.

Everything the bees make has prayer in it, even the scent of the hive itself. In the following pages, the bees reveal their deep commitment to the healing and evolution of the world and share ways in which we can assist them and benefit ourselves with their generous and remarkable medicines.

Honey: Kindling the Heart of Fellowship

As I'm writing this, we are in the middle of the honey flow, the time when the most nectar flowers are in bloom. In my area of the Pacific Northwest, the honey flow begins when the Himalayan blackberry blooms at the tail end of June. With their thick, prickly branches, blackberries are a bane to landowners, but their prolific flowers are a boon to bees. That bloom is quickly followed by fields of clover and, in my area, sunflowers. In other parts of North America, the honey flow begins when major bloom comes forth in alfalfa, locust, citrus, sumac, mesquite, loosestrife, and bird's-foot trefoil.

Two days ago, I gathered honey from a hive. The bees were gentle and sweet, floating up around me as I lifted off the hive box. Not a worried bee among them. I'm ever so careful not to spill a drop because that can incite robbing by the bees of other hives, who come upon a lick and wonder what other treats might be nearby. When I take honey, it is my responsibility not to endanger the hive in any way.

Hive boxes full of honey grace my kitchen table and sit on rimmed baking sheets on the counters. Last night we started processing the honey. We cut combs off the bars, and my husband pressed the wax with a potato masher to release the honey from the cells. Then he let it drip through a sieve into the five-gallon bucket that has a handy spigot on the bottom for pouring into jars. It's sticky business.

Because I know I'm going to get honey all over, I pull my hair back into a ponytail, put on a short-sleeved shirt, and apply some already dripped honey on my face. I worry less about getting honey on me if I begin by putting it there intentionally, and I get the added benefit of coming out of the sticky bottling process with lovely soft skin. Honey is a humectant and brings heavenly moisture into your skin. Using this natural bee gift on my skin is far easier than buying a lotion with a dozen strange chemicals—and is probably more effective, too. If you haven't experienced a honey facial, please try it. Man, woman, or child, your skin will thank you for it.

I save honey from my bees. *Hoard* is actually the more appropriate word, because even though I harvest it, I don't sell it. I hoard honey in case it is unexpectedly needed to feed a hungry hive. Because I never know when it will be needed, I like to keep two years' worth in my pantry. A few years ago, I acquired a hive in late November from a felled tree. When the tree hit the ground, the space where the bees had stored their honey smashed. All the honey was lost. The bees had no winter food and nowhere to gather any from.

I brought that bereft feral hive home to see if we could save them. My neighbor brought his front-end loader, and he and Joseph set into place the section of the old hollow tree that had survived the fall. We cobbled together their old home the best we could, sawing off jagged parts and blocking up holes. Joseph made a wooden feeder box that matched the interior shape of the top hole, and he attached the new feeding area at the top, where the hive's honey would have been stored. Over the winter and spring, I gave this hive all the honey I had. And I am glad I did. The bees survived and have been a strong colony ever since.

I suggest that if you keep bees, be kind when taking honey. Leave plenty for the bees, and save extra in case you have an emergency. Most beekeepers take honey in late summer or early fall, when the season is winding down. I take my cue from old farmer wisdom, which says honey is best taken in early summer. At this time of year, we know the bees have made it through the winter and have plenty of blooming summer weeks in which to store more. A good idea is to save a few bars with honey still on the comb so you can add them to a hive that needs just a bit more to comfortably make it through the winter. If you keep honey and honeycomb on hand, the issue of whether to feed sugar to your bees won't arise. If you run short of your own honey, ask a treatment-free beekeeper if you can buy honey from their hives. Or see what your local organic store is selling and ask if that honey is treatment free. If it is, stock up.

I eat honey, and I believe bees ought to eat honey, too. Some bee-keepers, in an effort to maximize profits, take honey from the bees and feed them sugar water or (egad!) high fructose corn syrup. They are taught to take more honey than is good for the hive, figuring they can always feed back sugar to make up the difference.

When their digestion is compromised by poor nutrition, bees suffer. A good diet made up of the foods they naturally eat makes a huge difference to bees. I am a stickler about honey, because it is what bees have evolved on. The pH, enzymes, essential oils, vitamins, and minerals are in the correct balance in honey, and that does not happen in any other medium. Bees cannot fully process foods that are foreign to them, just as we humans cannot eat grass. Everything I do, including feeding my bees, is done with awareness of how my actions affect them. If I ask the question, "What is best for these bees to eat?" the only answer I come up with is "honey." I can't pretend I don't know what is best for bee health. The bees themselves have this to say: "Sugar syrup is too high pitched, like treacle. We eat it if we're hungry, but it makes our stomachs hurt and makes our singing weak and tinny. Sugar doesn't have the prayer in the food."

I was deeply moved when I heard them say that sugar doesn't have the prayer in it; I realized that prayer is a necessary part of what nourishes them. When they process nectar into honey, the entire hive song coalesces into the prayer that blesses their food and brings them true nutrition.

Oh, blessed honey! Rudolf Steiner said of it:

> Beekeeping advances civilization because it makes man strong. Nothing is better for man than to add a little honey, in right measure, to his food. The bees, in a wonderful way, give man what he needs for the work of his soul. When he adds honey to his food he prepares his soul to work properly in his body. Bees ensure that man receives what is right for him. When one sees a hive of bees, one should say to oneself with awe and reverence, "By way of the beehive the whole universe flows into man and makes us good, capable people."

Honey is beneficial not only for bees but also for humans. Raw, unprocessed honey is full of vitamins, antioxidants, enzymes, flavonoids, and minerals. Honey that's raw and unheated retains more of its beneficial qualities. If its beneficial qualities are destroyed by heat, why would anyone want to heat honey? That's a curious story.

Honey's taste and color are determined by the flower source. In general, spring honeys, like those made from maple flowers, are light colored. Summer honeys are more amber, and fall honeys tend to be darker, though each of these seasonal assessments has many exceptions. Yet most of the honey you find in stores is the same midrange amber color.

A good friend who knows that I keep bees offered me a few gal-
lons of rich, molasses-tasting buckwheat honey and some crystallized
wildflower honey she got from an elderly beekeeping neighbor who'd
since passed away. She had never seen brown or crystallized honey and
was sure it had gone bad. I explained to her that buckwheat honey
is always dark brown, that crystallization is a natural process in raw
honey, and that honey doesn't ever go bad. Honey is a natural preservative, and even when thousands of years old, it continues to be edible.
Imagine that—a food from ancient times that you can still eat! Like
the hive itself, honey has the ability to live forever. Nonetheless, my
friend gave me the jars because she said her family wouldn't eat crystallized brown honey.

This is what marketing has done: it has created the assumption that
all honey should look and taste the same. To remove differences in
taste and color, the honeys on store shelves are heated and filtered to
keep them from crystallizing, and all the different varieties are mixed
together to produce a muddled, midrange neutral color and flavor.

Raw honey will crystallize over time, depending on which flower the
nectar came from and, to a smaller extent, how it was stored. Fireweed
and goldenrod honeys crystallize in as little as a week. Chestnut blos-
som honey and borage honey may stay liquid for years.

When the aim is to keep honey in its most natural raw state, it is
only lightly strained, to get bits of wax or bee wings out, and then it
is bottled. Pasteurizing (heating) and ultrafiltering honey is a more
unnatural process. The extra processing prevents the honey from crystallizing, which means it looks more uniform on store shelves. The
downside is that it is less nutritious. If you want honey's full benefits,
stick to buying raw honey.

Steiner said that honey was so precious that no adequate price could
possibly be placed upon it. In his "Nine Lectures on Bees," given in
1923, Steiner said, "Like bees, humans need nourishment that carries
over into our bodies. . . . If you eat honey, you will take into yourselves
a tremendous strengthening force."

Manuka honey from the New Zealand melaleuca tree has proven
antibacterial qualities, and I have heard of other parts of the world
where the proper combination of propolis elements give medicinal
qualities to other local honeys. Manuka honey is especially good for

wounds, and some research shows it may handle antibiotic-resistant staph bacteria as well. Manuka honey isn't the tastiest honey. It is more of a medicinal honey and is even used in hospitals in healing poultices for severe burns, bedsores, and skin ulcers.

As a biodynamic farmer, I am on the front lines of the clean agriculture revolution. We are thoughtful about what we purchase and everything we put in or on our bodies. Biodynamic agricultural practices encourage an additional spiritual component: treating everything with love and respect. Joseph and I put prayer into every step of our farming, raising our food to bring it to a state where it provides both physical and spiritual nourishment. I see in the bees their intention that everything around them be bettered by their presence, and I like to think that people are likewise inclined—that we seek to create a just world, full of generosity and love.

IN OUR OWN WORDS

Each drop of honey contains the rising helix, which invigorates humankind through its spiritual forces. Honey filled with spiritual forces kindles the heart of fellowship. Through this bond, humans develop respect and love of all beings. The bond of respect for all beings is the core of the growing heart. Through respect, appreciation, and all actions based on an honorable relation, you step into your evolution.

The taste of honey on the tongue is a lightness, a quickening, a deep emanation of the sun's light as all of creation bears witness and answers. One's heart is warmed. We bring this gift and ask that you absorb it into your being with awareness of the generosity with which we offer the fruit of a constantly creating and interrelating world. In this, bee and human become co-creative forces and light workers in the world and in love.

The Sting: The Fire of Love

These days, I rarely wear a bee suit to protect me from the bees. My bees are friendly, and I am familiar to them. I pay much attention to how they are when I'm around them, and if they seem concerned or upset, I move more slowly. Nearly always, my bees are sweet and gentle, and I can do most anything around them, even pick them up in my bare hands, without anyone getting upset.

Recently I harvested some honey from a hive in our bee gazebo. After I got the honey down to the house, I realized about a quarter of what I had harvested was still uncapped watery nectar and hadn't yet been dehydrated into thick honey. Nectar cannot be kept at room temperature like honey because it ferments easily and turns itself into the honey wine called mead.

I decided to give the uncapped honey back to the bees. I opened the hive for the second time that day and put two bars of uncapped honey back on the hive.

Normally I don't open a hive twice in a day, and I wasn't surprised that the bees were concerned. I got buzzed and then bumped by a guard bee. Guard bees take on the task of fending off interlopers, which I surely was. The first warning was a loud, distinctively cranky buzz near my face, calling my attention. A moment later the guard flew straight at me and bumped me, forehead to forehead—her way of saying, "Back off!"

An irritated guard bee has no regard for scale. Even though I'm gigantic compared to her, she still flew right into me and tried to push me back. When a bee bumps you, you have a scant second to take the hint and back up. If you back up ten feet, she'll probably leave you alone. If you don't, the next line of defense is a sting. I took the hint and backed up, and no damage was done. She warned me, and I did as she asked. A truce.

That night, I looked at all I'd harvested and decided I didn't need to keep as much as I had taken and could easily give two more bars back. The following morning all the hives looked quiet and happy, bees flying in and out their front entrances. My friend Susan offered to help, and together we opened up the hive and put those two bars of honey back in.

A young colony in the top-bar hive next door had a recent growth spurt; so on a whim I decided to peek inside and see how the combs were coming. The normally calm, even hum from all the hives had the tiniest edge to it, just a quaver higher than normal, so I decided to wear a bee veil. I don't generally mind getting a sting now and then, but a cranky bee tangled in my hair is decidedly unpleasant for both the bee and me. Wearing my veil and being quiet as I could be, I removed the top of the hive.

Eighty bees came screaming out, and I got a half dozen stings on my hands in two seconds! I quickly backed up and put fifteen feet between me and the hive before I turned around. I then walked fifty more feet away, with bees still stinging their hind ends into my jacket and on the screen of my veil.

When bears are chased by bees, they find nearby low-hanging tree branches and move around in them. The waving branches confuse the bees, who lose sight of the bear as he hightails it out the other side of the branches. So that's what I did: I trotted over to a nearby peach tree, raised my arms into the branches, and spun around in a wide circle. Then I stepped out from the other side of the tree and listened to the air around me. Two bees were still angrily stinging my sleeve. I brushed the soon-to-be-dead guardians off into the grass and said, "Thank you for your service to the hive." Then all was quiet.

This colony had never, ever been feisty. I was completely surprised, wondering aloud what had caused the extreme upset. I reviewed how I'd handled the bees and found no error. But surely it was something. What could it have been?

Then Susan astutely pointed out it was the morning after Independence Day. Fireworks are legal in my state, and the evening before, fireworks had boomed into the wee hours. Many were set off by rural neighbors who live a few hundred feet from my hives. Egad! That must have been it. Joseph and I had been up in the field, finishing up animal chores just after dusk, and we both noticed how the ground shook from the nearby fireworks. Bees don't like that kind of vibration.

If you have plans to have a look in your hives the day after your bees have been violated by loudness, like mine had been, you may want to rethink that. I am so glad I'd listened to my intuition and put the veil on.

In case you don't know how to best remove a stinger from your skin when stung by a bee, here is how. The stinger is ribbed, so it stays in the skin after a sting. When the bee punctures your skin with her stinger, the stinger separates from her body. That maiden will die, because the back end of her abdomen came off with the stinger. The stinger, now free of the bee, continues to pump venom into the site for about thirty seconds.

Most people use a thumb and forefinger like tweezers to pinch the stinger and pull it out. Pinching like that will simply inject all the venom. A better way to do it is to use a single fingernail (your driver's license or library card will also work) and scrape the stinger out. That stops the venom pulse, so less venom gets in.

Some folks think they are allergic to bee stings, but most people are not. It is normal to have swelling from a sting—sometimes substantial—and it may take a few days until that area is back to normal. A true allergic reaction involves anaphylaxis that swells the throat enough to stop breathing. Only a tiny portion of the population has that response.

Back when I was a new beekeeper, I was afraid of stings. While I still don't enjoy getting stung, I don't mind it so much now when I do. I believe the bees know where we could use a bit of sting therapy.

Bee venom is a potent mix of anti-inflammatories with antibacterial and antiviral properties. It has been used in the healing arts since the time of Hippocrates, the "father of medicine," who prescribed bee-sting therapy as a solution to joint problems and arthritis. Bee venom has the ability to promote healing by stimulating a healthy immune system response throughout the body, especially at the sting site. The venom wakes up our ability to heal ourselves.

A few years ago in February, Joseph and I helped an elderly friend prune four hundred yards of grapevines on his farm. The next morning my thumb was so sore I could hardly fold my fingers together. All my fingers and especially my thumb ached.

Later that morning, I was up at the bee house checking on my bees. It was winter, so no bees were around, but while I stood there, one bee came out for a cleansing flight. At the entrance of the hive, she readied herself for lift off, but then she turned and flew to me instead, landing on my wrist. I was in the midst of another wonderful bee moment when she stepped onto the heel of my hand at the base of my thumb, turned her face toward me, and while looking straight into my eyes, leaned back and stung me.

She certainly surprised me! Once you've been stung, there is not much reason to flick the bee off, since the stinger has already disconnected. I hadn't been stung in many months, so I let the venom pulse into me. Within a few short minutes, my thumb completely stopped aching.

Another time I had seven stitches over a tendon injury, and a bee stung me just below the bandage at the base of the seventh stitch. The cut healed astonishingly fast, with no residual soreness or disability.

I have had this experience with bees numerous times. Something in my body needs attention, and my typically gentle bees send an emissary on a mission to relieve my discomfort.

Every summer at our county fair, I volunteer at the bee booth, which is divided into two rooms. The display room on one side of a partition is filled with bee paraphernalia and blue-ribbon honey entries. On the other side is a wire-enclosed, ten-foot-square area with a live hive of bees inside.

Every half hour, a volunteer gives a talk about bees while standing in the bee room. He or she enters through a screened, double-door airlock. Most volunteers wear the bee suit. I want people to know how gentle bees are, so I do my talk without protection. Inevitably someone asks how that is possible, and I explain that bees have little desire to harm anyone. They sting only when they are worried that they or the hive are in danger. Though the number is higher now, back then I told the audience that I'd only been stung three times in my life, each an accidental sting when a bee got tangled in my hair or clothing and, fearing she was trapped, stung me. No harm meant—just a scared little bee.

One year at the fair, I was paired with a beekeeper who has political ideas opposite my own. The bee room is not, in my opinion, the right place to spout one's political beliefs, but that's what he did. I was a bit on edge (understatement), and I asked him to skip the political statements and stay on topic. But he still interspersed his views throughout his bee talk.

When it was my turn to go into the bee room for the demo, I was not in the heart space I usually am—and need to be—when I approach bees. Not even close. I stepped into the airlock to the bee room, and *wham,* I got stung right on the top of my head.

I stepped back out of the airlock and untangled the little bee from my hair. I had already pumped some snitty-ness into my system before I got stung, and I was surprised at the adrenaline the sting evoked in me. I shook it off, took a deep breath, and walked back in. *Wham!* Again I was stung on the very top of my head. The exact same place as before—my crown chakra—the side of me that points to the heavens, though I certainly did not have heavenly thoughts emanating from within.

I stepped out of the cage again, and this time a single bee came outside to have a talk with me. She assertively buzzed me and in no uncertain terms told me not to set foot in their home until I had worked out my "stuff" and bettered my attitude. Okay, I could take a hint. In utter humiliation, I went out and returned in my full bee suit, wearing it the whole time I gave my talks.

The relevance of receiving not one, but *two,* stings in the exact same spot was not lost on me. I admit I had very little loving cosmic consciousness going on when I first reached for the door to go inside the bee room. I was still replaying, "I should have said . . ." As I am not yet an enlightened being, I still had that dialogue going around in my head as I drove home. Once there I changed my clothes and decided to visit with my own bees to calm myself down. I walked barefoot through the field toward a hive, and *wham,* I stepped on a little bee, who stung me in the very center of my foot, on the Bubbling Spring acupuncture point—the very first acupuncture point that forms in the fetus, the one that helps you ground and connect with the earth, the point that roots energy downward.

I sat down on the ground, scraped the stinger out of my skin, and apologized to the honeybee for stepping on her, feeling terrible that I'd done that. I told the bee her gift wasn't in vain, that I would sit right there, go over my day, and let go of whatever crap I was carrying around that was making me bad bee company.

Three stings in one day doubled what had, until then, been my entire life's sting count. Getting stung on the very top and very bottom of my body was a none-too-subtle message. I had throbbing focus points showing me precisely where energy runs through and connects me with the heavens and the earth.

I sat there in the field and apologized to everyone I had labeled harshly, had miserly thoughts about, or had offended (including my higher self) by being such a knucklehead. When I felt I could be a better person, I got up and went on with my day.

I did notice that as a result of having all that bee venom in me, I was intensely aware of *all* of my body, like I was breathing through my skin instead of just my nose. Having a front-row seat to an important bee lesson in selflessness caught my attention, and I felt buzzingly happy for quite some time.

IN OUR OWN WORDS

The sting. I inject myself into you; when you sleep, you dream of possibility. Even the discomfort is to awaken you. I kiss to start you up. The turn is the fire of love.

My venom cures disease and purifies the blood by stimulating the cleansing process, but not by removing a substance. No, this cleansing is by transformation. We commingle our matter in a pulsing swelling that rises up like a small hive under your skin. Wild within the small cosmos, it calls your attention. Immerse yourself in this heightened awareness and feel the helix that is my signature.

I die into you that we both may live.

Evolutionary Medicine: Bathed in a Sacred Illumination

Bees are not simply a backyard hobby for me. They are intelligent, respectful, caring beings who are full of commitment and love. I am proud to call them my dear friends.

When I first started speaking with them, I was thrilled to understand the inner workings of the hive, how they interact, what they see as their role in agriculture, and how they envision their larger purpose. With all that, it hadn't occurred to me that they would also expect me to understand advanced concepts of healing and medicine and even quantum physics.

Some of what they have spoken to me has set me researching these topics just so I could comprehend their words and images and continue our conversation. I have sat at my computer, asking the Internet what a quark is and how the limbic system works. Over time I imagine the bees will bring many more concepts that will send me back to Google again and again. Occasionally in these web searches, I come upon information that verifies what the bees have told me, which is always exciting.

Rudolf Steiner also saw deeply into the potential of the bee to heal and to assist us in our own evolution. In his "Nine Lectures on Bees," he said:

> The consciousness of a beehive, not the individual
> bees, is of a very high nature. Humankind will attain
> the wisdom of such consciousness in the next major
> evolutionary stage when human beings will possess
> the consciousness necessary to construct things with
> a material they create within themselves.

Since 2010, when I first heard the words of the bees, I have, of necessity, expanded my own thinking. To get this bee knowledge out into the human world, I've had to become public about what the bees have said and acknowledge that this information has come to me from the bees themselves. The first few times I spoke of my communications with bees, self-consciousness made me hesitant, and I had to push past that self-imposed barrier. Yet the bees have been clear to me that they want us to know them in a newer, deeper way. They are always compassionate with my sometimes bumbling efforts to understand them better and to serve as their voice.

In one conversation, they explained how they can help humans become healthier and how they heal disease. They are eager to help us progress into our next evolution and become more caring and connected in a vital way with each other and with our precious Earth.

I began this communication by asking them where bees came from.

IN OUR OWN WORDS

We are an evolution unto ourselves. We emerged into the earth sphere as a template of unity, harmony, community. Bees are an evolutionary progression.

Scent and sound have the ability to enter the nervous system to provide a progressive adaptation to the environment. The scent and sound of a hive contain compressed information. Scent and sound can rewire our deep inner perception—the matrix of what we operate from, our belief system. This is no small task.

Each hive's great hall is the portal that allows each hive to be present in the consciousness of all hives. Bathed in a sacred illumination, our communal presence kindles the evolutionary intelligence nascent in each being. And thus we begin.

We prepare and build our hive to contain the scent. The air we breathe inside the hive is our medicine as well as our joy. We seal the hive to keep harmful influences outside, to preserve our hygiene, and to magnify the forces impregnated in the interior air.

Air is a canvas. Like water, it has memory. As we breathe, work, and imagine, we imprint the air within our hives with industry, unity, solace, and love. The Unity of the hive is expressed in our sound, which permeates the air. The air within the hive is full of our song, full of the event of us as we fill time forward and back.

We impress the noble scents of nectar, wax, propolis, our queen, the emotional states of all the bees. Yes, we have emotions, but not as separation between us. We feel as one in our individual contributions. The joy of the harvest tasks, the delight in the scent's vapors, the keening in the loss of our queen—we suffer or celebrate together.

A wild hive is sealed, and the colony lives in the propolis-scented hive air, breathing medicines that keep us healthy. Our systems respond to the enumerated quavers by bringing each bee's body to balance and good strength. A hive in good order lives in the healing scents of the medicine we have made for ourselves from plant partners who surround us. This medicine offered by the plant community includes the rising mineral-rich sap of trees, their lifeblood.

Humans will eventually begin to use scent more to heal their bodies and minds. Again, this is medicine by addition rather than subtraction. Subtractive medicine seeks to kill an organism or relationship, but it often has deleterious side effects that further unbalance the organism's health. Medicine by addition restores balance by completing relationships, like keys in locks. In its highest state, medicine by addition further opens beings to optimal states, creating evolutionary expansions that resolve past weaknesses. These advances include the physical states, of course, but they also direct us toward advanced and enhanced intellectual, emotional, perceptual states that often combine and overlap new abilities.

When humans begin participating in medicine by addition, they will strengthen their systems and increase their sensitivity, vitality, and perceptual abilities, which sets up the conditions through which evolutionary progression occurs.

Quantum Healing: A Temple of Unity

My dear friend Michael Joshin Thiele says, "Altruism is one of the Bien's deepest life gestures." I love that. It matches completely my experience of the bees.

When I spoke with the bees about their medicine, they offered to share a wondrous healing method with humankind. They are exceedingly generous, full of goodwill, happy to share this process that supports human health. I hope you use it with reverence and respect, in the manner they describe. They first showed me an image of many bees opening the ribs of their hive-body to expose their heart, giving us trusting entry.

As the Unity, the hive is both body and being. Used erroneously, this method can cause them great harm; so please, if you do this, maintain the utmost integrity in your actions.

I have benefited from this prayerful healing system for some time. I have also used it with others who have health issues.

A good friend has been significantly helped by this method. Her nervous system and organs had been damaged by a poisonous tropical spider bite four years earlier. She suffered from dizziness, pain, headaches, organ pain, and visual disturbances. She was so ill from the ongoing neurotoxic aftereffects and organ damage that she wondered if she'd survive. After engaging in this healing process for half an hour, she said her symptoms were greatly reduced, and successive work with the bees has taken her in a very positive direction. Over time, she has incorporated this method into her daily routine with many beneficial results.

Sitting at the entrance door to any hive is the best place to do this. If you have bees or can find a beekeeper who will allow you to sit quietly near a hive, you may do it right next to the hive in the warm months of the year. If being in the company of bees is not practical, or it is during the cold months of winter, the bees offer another way to unite the three aspects of this process—prayer, sound, and scent—in your own home. *Note:* The resource section at the end of the book lists materials you can order, such as CDs, downloadable recordings of the sound of a thriving beehive, and fragrant, healing propolis.

The following are the bees' description of the healing process, plus my additional instructions. This healing method may look simple, but it is very powerful.

IN OUR OWN WORDS

We tell you of a way that bees can heal humans. Enter into prayer with us as we offer the gifts of scent and sound. To partake of our medicine, do this in an atmosphere that recognizes the gift, and in a state of reverence. In this manner, we of the bee kingdom offer humans a template to create a temple of unity.

In our healing, we offer a multidimensional path to open your mind and body to healing, and your heart to love. We offer humankind a way to evolve to a place of harmony and supportive community in our shared environment on earth. In this process, we have created medicine that is a healing balm for your bodies and a process that engages our souls in evolutionary progress.

Prayer

First there is prayer. Ask for healing. When you come in prayer, you open yourself to a state of grace. Let generosity flow from your heart. Come with love. Share with generosity.

This process is an exchange, not a taking. Transformation and evolution take place in the arms of joy, generosity, compassion, kindness, love. These grace states elevate our being, opening and making available the healing and evolutionary access points.

Step 1: Sit quietly by the side of a busy, thriving hive. (Don't block the flight path by sitting in front of the entrance.) If the bees continue their activities with little regard for your presence, consider yourself accepted and welcomed.

If sitting by a hive is not possible or the bees have withdrawn into the hive for the cold months, find a place in your home or yard where you can sit quietly and bring yourself into reflective, open mind space. Bring your heart in fullness as a gift. You must not simply take from the bees. Create a heartfelt energetic meeting, with love and joy freely given. Open yourself in a prayerful and contemplative manner. Grace comes in many forms, such as gratitude, peace, love, joy, inspiration, and beauty. Let the feeling fill your body. Take time to immerse yourself. Do not chase it. Allow it to come into you.

Sound

Step 2: Bring your awareness to the sound of the hive. If you cannot sit at the side of a beehive, listen to a recording of a healthy hive through headphones. Immerse yourself in the sound.

Listen to the sound we make. Let it fill you and bring you into our healing space.

Propolis Air

Step 3: Just as you focused your attention on the sound of the hive, now put it gently on its scent.

Our hive air is filled with the healing elements of the propolis. Propolis is medicine. It has the ability to go into every cell and bring healing, especially when it is airborne. It can be ingested, but it is even more powerful airborne. The propolis vapors contained in the air go directly into the blood. This stimulates an energetic state that contributes to our healing and our evolution. This can alter, repair, and build anew our DNA. This is not about sniffing bee air; it is a sacred time to be present in generosity with love, and out of that comes joy, the communion of our sharing.

Propolis contains the building blocks that strengthen cells, promote healing, and bring about an evolution by communicating new information to our cells. Evolutionary medicine addresses damaged cells and heals them. All the components are present. The different combinations within propolis and within the hive are tapped.

If you are sitting by a hive, simply breathe. In warm weather, the air around a hive is infused with the scents of honey, propolis, and the bees themselves.

Cold air is not infused with scent, and in cooler months, the bees will have withdrawn into the hive, where they put all their energy into keeping themselves and their brood at a precise temperature. The inside hive air is meant for the bees and must be respected. Disrupting this balance can do great harm to the bees; so if it is cold where you are, do the following step at home instead: Put a thumbnail-size ball of propolis in a small, covered glass jar or other confined air space. Leave the glass covered in a warm room for a few hours. This allows the scents to accumulate and magnify their purpose. In the warmth, the propolis releases its healing vapors inside the jar.

After a few hours, remove the lid and breathe the propolized air slowly until the breath is full. Propolis is best taken into the body by inhalation, which allows it to rapidly enter the bloodstream and nervous system.

In both situations, near a hive or in your home, gently hold your breath until the scent of the propolis fully absorbs into your body. Continue breathing the scented, infused air. Breathe deeply but normally (don't hyperventilate) for at least ten minutes and up to thirty minutes.

The scents are keys that fit into matched locks. Once matched and filled with a key, the lock opens. Stay in a receptive and meditative state while breathing the air into you.

 Here is the alchemical combination:
- Propolized air
- Harmonized sound
- Bringing oneself to be fully present in healing
- Knowing how to bring forth unity
- Being in the grace states of love, generosity, and joy
- Having the courage to advance in gratitude and with respect for all life

Hive air also contains our exhalation—each little bee releasing what came before. Thus, our healing moves us ever forward in the loop of time, one of the many possibilities we lay our experiences upon. Through our breath, we are placed within this moment and all it offers us for growth. As we breathe in, we take in information about our environment, our relationships with all that is around us. What we find either nourishes or depletes us. Through our breath, we continually strengthen our bonds to that which we imagine—to stress, fear, and worry, or to love, joy, and kindness. Each breath in, each breath out.

The breath in informs us of the world we dwell in. The breath out is our response to that world.

Thus, it is the same with bees and humans. When we are stressed or worried, we fill the air around us with that concern, and healing may be difficult. To heal ourselves we may choose differently. What is introduced into our being has the ability to harm or heal. Our breathing—such a simple act—brings us to each moment wherein we imprint that choice.

We are, after all, God's own question seeking an answer. What choice will we make, and how will we live with that? Whether we are in communal group thought or the thought that comes to us alone, we carry forward our experience and responses to what we know.

To maintain our healthy state, we seek a clean environment. We work at the door to guard our perimeter to keep our home free of that which causes concern. If something slips in, we do our best to isolate it, even encapsulate it. In this way, a threat to our health is moved from the interior (our home) to the exterior, where it cannot harm us.

Hive air moves freely in and around us. Our individual and shared responses to the air strengthen or weaken us. With each individual and collective breath, we confirm our response. By that, we live or die.

In the healthy hive, the air is capable of dismantling attacks and preserving health. We become strong in each moment and capable of doing our good work. Even in the presence of that which can cause disease, the medicine built into our home allows us to flourish.

Look what you surround yourself with. See what you bring into your home and ask if this brings you peace or whether it causes you harm, even a little. If it causes harm, clean it out. Once your home is secure and supports your health, you can enter into the larger environment with the wisdom to do the same outside.

Your breath brings you daily strength. The world comes into you, mingles with your body and your thoughts, and you exhale air that expresses the being that you are.

If love circulates in your being, that which you speak and feel will come out of you. Even if what you bear is difficult, the peace you carry within will also carry out into the world. Thus, all your actions are informed by love, and you become an instrument of peace and kindness.

Look within your own home. Build kindness into your relationships and care for the environment as if it were the interior of your own home, your own body. Your home, then, becomes powerfully supportive medicine that heals your weakness, gives you strength, and leads you to create that which is beautiful.

Visionary Bees: Realm of All Possibility

This chapter came to me while I was on a writing residency at Hedgebrook, attending a women's writing retreat. Early one morning, I had been wondering about possibilities and thinking about the vast number of situations I'd hoped for in my life—like this residency—and the good luck of having so many come to fruition. I was reviewing the sequence of coincidences and events that had led me to Hedgebrook, recalling each step that had to occur before the next to keep opening doors, until I eventually arrived here, in the forest on a spring morning on Whidbey Island.

It was easy to see how some of the events, like submitting my application, led me here. Others seemed more random, such as my sister in Southern California reading an article about a writers' retreat in Washington state and suggesting I inquire about it from my old home in Massachusetts. The most critical step was my questioning the possibility of getting accepted and resolving to try. I could have easily decided it was too hard or too much effort, and the story would have stopped there. But then it all fell together, and each thought and every small action ("Do I have enough stamps to mail my manuscript?") was answered with a yes.

I pondered other serendipitous situations that led to unexpected and positive results, like the summer after college when I rented a tiny cabin on Martha's Vineyard and took my first job teaching. One day, while sitting outside the cabin, I decided to dig a hole in the side yard, sink a tub in it, and make a little pond with the hope that frogs would find it. My childhood home had been on a pond, and I longed to hear again the sound of the frog chorus I'd fallen asleep to each night as a child. I borrowed a shovel and started digging. A few feet down I hit a big rock, and I couldn't find the edge of it. I scraped back the dirt from the two-foot deep hole I had dug and saw white enamel. I was digging out dirt from inside a sunken tub, just where I'd imagined one should go. I finished digging, put rocks, plants, and water in the tub, and within a week, frogs found it and began singing the song I'd longed to hear.

With all that on my mind, I asked the bees what I thought was a non-bee-related question: How does this—bringing ideas to fruition, marvelous coincidences—all work? Here's how they answered.

IN OUR OWN WORDS

The quantum field is the envelope of God's thought. The field is permeable and easily accessible. It is the realm of possibility, the place of imagining. The quark comes into existence with thought as we create ideas. Multiple versions easily occur as we come to agreement on which of them to follow.

In this realm of all possibility, diamonds may be born out of a thought of coal. What we imagine can be so.

These ideas bubble up within the medium, hovering in the matrix. The conceptual framework of thought exists in many dimensions. As such, the image of bees popping through the field into a relationship with humankind was just such an event.

One of the many ways to create a quark is to imagine the future. One could just as easily imagine the past. No need to get caught in illusory constructs of time; it's all within the realm of possibility. What is not, soon is, though it is challenging to talk of the field outside the hem of time. It is only the construct of a language formed of past, present, and future that wraps our thoughts. All is possible. Ideas based in time are as simple as Möbius strips.

Our development has come through understanding mechanics and cause and effect—a wonderful evolutionary step—that delight the intellect. We create a quark and place it into the field where it bakes like a pie.

Time * Space * Emotion (fuel) * Possibility

Bees move easily in and through the quantum field. Location is a good example. A bee scouts a new home as the swarm rests on a branch. When she finds a possible home, she enters it. As she moves through the room, her mind overlaps the template of "hive" onto the space—projecting it like a movie, fully inhabited, knowing how the colony will fit and function here. She draws the image from the field where the template already exists. As she does this, the projection simultaneously appears—knowing in the consciousness of the swarm, who is in a different location. Thus, the swarm knows the suitability of the site before they have even visited it.

When a bee makes the exciting discovery of a tree in bloom, she puts emotion into a waggle dance that communicates her joy to the other bees. Her dance also gives directions to the tree. Her joy-filled dance helps "make it so"—helps connect the other bees with the tree. Emotion fuels the manifestation. The dance sends creative energy to the quark and helps it manifest from idea to reality.

The dance is a practical way of communicating a location, but the emotion and excitement of her discovery help turn a distant tree into a source of food—the bee's original intention. The dance is the fuel, just as all hope, desire, loss, joy, anticipation, frustration, exuberance, worry, doubt, delight, and other emotions are. Each emotion has a distinct fuel that aids manifestation. Love is also fuel, but it is different from these other emotions; it is part of the fabric of all that calls us into being. Love is the medium upon which all ideas birth. The quantum field itself is love. From here springs all possibility and our ever-emerging, advancing evolution.

VII

The Song of Sharing

How We Can Help Our Bee Friends

Bees have been serving us for thousands of years. Now they need our help. Whether you have hives of your own or just care about bees and would like to be as bee-friendly as your circumstances allow, you can help bees in some way.

If you have a yard or even a small balcony, you can plant bee-friendly forage for your own and neighborhood bees. You can also provide a friendly water source for bees. In dry weather, a bee colony sends groups of bees out to find water and carry it back to the hive for the bees to drink and to keep the proper humidity in the brood chamber. My favorite bee watering stations are birdbaths, but you do need to modify them for bees. Make a mound of stones, gravel, moss, and wood pieces in the water so bees can safely walk to the water's edge without falling in. Be creative! My friends Robin and Jody build the most beautiful bee watering stations by filling concrete birdbaths with crystals and moss. Place the birdbath in the shade so the sun doesn't evaporate it, and make part of your morning routine a pleasant walk to the bee watering station to refill it. You'll be surprised how much life—butterflies, ladybugs, dragonflies, praying mantises, lacewings, and other creatures—will center itself on this station.

Perhaps you would like a hive of your own. If you have the room and the inclination, keeping bees can be immensely rewarding when done with respect, love, and kindness. There are beekeeping groups in just about every community; look for one that practices natural and treatment-free beekeeping. If you already have hives, you may want to learn more bee-friendly practices so you can be an even better friend and guardian to your bees. Join others who are committed to bee-centric beekeeping and support each other in learning best how to do this. My key questions whenever I am about to do almost anything with my bees is, "Are my actions in service to the Bien?" I ask that so I seriously look at how my decisions affect not only the singular hive I'm standing next to, but also the Bien of the world. It is my sincere hope that the lessons the bees share in this book will influence many beekeepers to use their principles and become bee-centric beekeepers.

If you aren't prepared to care for honeybees yourself, perhaps you would like to offer a beekeeper space for a hive or two or ten on your land. I know many beekeepers who would like to have more hives than their land can sustain and are always looking for new sites for new hives. This way, you can enjoy bees and leave the keeping to someone more knowledgeable.

Can you write? Send a letter or email to your government representatives and let them know that you support (with your dollars and your vote) bee-friendly legislation. Let pesticide companies know that you will not purchase anything they make that harms bees. Talk to your friends about bees and what you have learned about them. Donate to the Xerces Society for Insect Conservation and other organizations that work to keep the world friendly to bees.

Eat organic and non-GMO foods. Gene-modified foods are designed to be resistant to pesticides and poisons that are harmful to bees and other life. When GMO fields are sprayed with poisons and planted with chemically treated seeds, much of the vital life force in the area dies. Please don't support this industry. Organic practices that support the growth and interdependency of the land and its inhabitants are a far wiser way to care for the earth.

The bees say honey is food made in prayer. I am convinced that honey truly is spiritual food. Honeybees are created by the hand of God, and I believe in them and their mission. I feel blessed when I have a spoonful of honey in me.

I'd love to see more beekeepers question the wisdom of conventional beekeeping methods and move to bee-centric, clean, treatment-free systems. It starts with us. To build more awareness of these ways, ask the beekeepers at your local farmers market how they care for their bees, and let them know there are alternatives available. I suggest bee-friendly ideas and practices when I speak with beekeepers who don't know there are alternatives. Have a look at the resource section for sources of information on bee-centric beekeeping.

Most of all, you who are beekeepers can take a stand for all the bees you meet. Raise the healthiest bees possible, free of poisons and deeply engaged in their local region. If you must buy bees, buy them from local sources so they are familiar with your weather and vegetation and hold within them solutions to regional health issues. Be in service to Mother Nature and provide the healthiest environment you can so the Bien may grow and develop in accord with its own wisdom.

Planting for Bees:
Light Anchored in the Ground

In early spring I set myself to gardening. March is cold in my part of the world, but violets, primroses, daffodils, and cherry trees are blooming. They come fast on the heels of the cold-hardy snowdrops, hyacinths, and pussy willows. Even though the days are nowhere near balmy, if the sun peeks out, the bees fly out seeking early nectars and pollens.

Bees always surprise me with their commitment. If it's sunny and marginally warm—even on days I reach for my hat and gloves—they will be out harvesting from the blooms. On gray days, if the sun reaches their front porch, and the weather breaks for as little as fifteen minutes, I see them dash outside. They measure time by whether they have long enough to gather a few pollen specks and a tiny drip of nectar before sprinting back to the warmth of the hive. If it is, they go.

Bees need plenty of flowers, and the closer those flowers are to the hive, the more time they can spend gathering pollen and nectar and the less time flying. Whether you keep bees or just want to help bees, plant flowers. Here are a few guidelines to help you choose what, when, and where to plant for bees and other pollinators.

Plant Flowers in Bee-Friendly Colors

Bees are more attracted to certain colors than others. Bees prefer, in this order of priority; purple, violet, blue, blue-green, yellow, and white. Keep those colors in mind when you are planting your bee garden. Bees have a hard time seeing red (it looks like black to them), though they can see orange. Many colors visible to bees are not in the same spectrum as colors for humans, which is difficult for us to imagine. Bees perceive more through the ultraviolet (UV) spectrum, which also helps them see where the nectar and pollen are hidden.

Plant Flowers Close Together

Bees like to forage in areas that have plenty of flowers, especially similar flowers, so group your plants into a smaller space rather than separating them into distant areas.

Sometimes people tell me, with great sadness, that no bees visit their garden. I have a few ideas about what may be causing that, and a solution for each.

1. If your neighbors don't have blooming flower gardens, it may not be worth the bee's time to fly so far for one garden stopover. You can attract them by making your place a spectacular pollinator kingdom.

2. The flowers you have planted may be scattered too far apart and would take too long to pollinate. Make your plantings more cohesive.

3. You may live in a neighborhood where people use bee-killing chemicals. Create and nurture a chemical-free environment in your own yard and encourage your neighbors to do the same. Some people have even gotten their local communities to stop using toxic chemicals in parks and roadways.

Plant Single-Petal Flowers

Bees don't want to spend valuable time figuring out where the pollen is on a multipetaled flower. The pollen on single-petal flowers, like cosmos and sunflowers, is easier to find than complicated chrysanthemums and tight-petaled roses. Stay simple.

Flowers need to have nectar and pollen openings that match the bee's tongue length and body size. The throat of monarda (bee balm) fits a bumblebee perfectly but is too long and narrow for the honeybee's tongue. Squash flowers are wide inside and a riot of joy for honeybees. I've seen as many as five bees inside an open squash flower at once, all of them covered head-to-toe with golden pollen.

I dug out a clump of flashy Asian lilies when I realized none of my bees had any interest in them. Our honeybees are originally European honeybees, and they do not have a historical memory for flowers from an unknown continent. I planted bee-friendly rudbeckia (brown-eyed Susans) instead, and they're well visited by bees.

Plant in Clumps

When a bee visits a garden, she is on a mission to gather pollen from one kind of flower. That is what makes pollination work. If your sage, mint, lavender, and wisteria are all in bloom, one group of foragers will focus on lavender, while another visits the wisteria. Rather than planting lavender here and there, put all your lavender plants in one close area. The bees will be happier and can be more efficient.

Provide Blooms in Every Season

In the Pacific Northwest, most of our blooms happen in spring. Everything flowers until the drier weather of midsummer comes. Less and less comes into bloom as we move toward fall. I want my bees to be busy right to the tail end of October, so I plant lots of summer- and fall-blooming plants, like Joe-pye weed, goldenrod, asters, salvia, and autumn joy sedum. Wherever you are, be sure your garden has a few things in bloom through all the seasons your bees are in the garden, and plant extra in the slower seasons. Don't let the bees run out of food.

Provide Bee Medicine

Plant flowering herbs. Not only do bees collect nectar and pollen from herbs, they also gather essential oils from the flowers. These essential oils are added to wax and propolis to create the air-infused medicine the bees breathe. Lavender, borage, catnip, fennel, all the mints, rosemary, sage, and thyme are particularly good for bees.

Use No Poisons

If your neighbors are using bee-killing poisons, it will be difficult to attract bees and keep them alive. They will travel all through the neighborhood, and sooner or later, they will find the poison. My experience is that most people are quite unaware that our laws still allow bee-killing chemicals on the market. Many people still believe if they can buy it off the shelf, it is safe enough. Educate your neighbors about organic gardening so bees can survive.

I have been working for years to persuade neighbors who use toxic poisons to shift over to organics. It doesn't usually happen overnight, but with gentle persistence and friendly help, it can be done.

A pesticide kill is an egregious crime against nature. Over the years, I have lost hives three different times from someone using pesticides nearby. Because bees travel up to two miles from their hive in their pollination duties, I was hard-pressed to figure out who had sprayed pesticides on a given day. I do, however, know that it happened the same way each time.

The about-to-become-a-bee-killer wakes up on a Saturday morning and notices his peach tree is blooming. He suddenly remembers he forgot to spray the tree when it was still in bud. He ignores the directions that say never to spray when a tree is blooming, because that's when pollinators visit. He disregards that the directions say to spray only on windless days so the poison doesn't travel far, and that it's best to spray at dawn or dusk when the bees are not out. To the person spraying, it's midday on a weekend, and he's got time right now. Better late than never. Not noticing my bees in the tree doing their pollinating job, he sprays the tree, and the bees unknowingly bring that poison home to the hive.

Bees touch each other numerous times throughout the day, especially when another bee is having difficulty. One study said that if a handful of bees come home sticky with poison, within twenty-four hours, every bee in the hive will have touched a bee who touched a poisoned bee, and the entire colony will be poisoned. Each time one of my colonies got poisoned, they died a slow, painful death over twenty-four hours. I watched with tears in my eyes as they succumbed to the terrible neurotoxic effects, and there is no doubt in my mind they suffer badly. There was nothing I could do to help. Once the poison is in them, they are on the death path. Joseph dug a hole; we buried the bees and burned the hive.

Recently I was very worried about a neighbor who, each year, sprayed toxic chemicals on her fruit trees. At first she was not open to changing how she gardened, and I struggled to find a shared language that respected her right to garden as she wanted and my bees' right to live in good health. I helped her with small projects, offered to run errands on my way to town, and shared garden produce. Eventually I asked her if she'd let me know the day before she sprayed so I could keep my bees inside. This year she called me the night before, and I thanked her for her consideration.

I then asked if she'd tell me the name of the pesticide she'd be spraying. After going online to find out what it did, I called back to ask if I could buy her the organic version. I even offered to pay for it because the alternative—many dead hives—was so horrific. She agreed; so I drove to the nearest nursery that stocked the organic version and bought it for her. She used it that night, and my bees survived.

Keep in mind that poison is poison. Even though something is organic, it doesn't always mean it is bee friendly. I helped my neighbor read the directions and encouraged her to spray on a windless day toward evening when my bees were all safely in their hives.

I encourage people to use their common sense and examine why they would want to eat food that has poisons on it in the first place. Poisons kill—that's their purpose. How can that not have an effect on our own bodies, our children's health, and the longevity of honeybees? We can start making changes by becoming sensitive to our own health and avoiding poisons. From there, we can extend that attitude out to bees and bee health.

IN OUR OWN WORDS

What makes bees happy? The shape of scents, the sweet of nectars, the sun's prismatic light, our cumulative joy. In almond orchards or places where monocultural practices hold sway over nature, we miss the natural world and suffer in our longing. We would benefit from beds of herbs planted where a tree has died. We ask you to make garden beds and fill them with our medicine: flowering herbs.

Flowers are an embodiment of light anchored in the ground. The plant has a relationship with the sun and the light that raises it toward the sky and sun. We go out and collect pollen, which is materialized light, and bring it back to the hive.

Each grain of pollen is an anthology of information about local flora and the mineral terrain. We develop and grow on a diet of the landscape. Each mineral and plant speaks to us in our nascent form, describing in a wordless language the pitch and flow of the flora.

Later, after we are born and flying, we experience the jubilant excitement of discovering these flavor-scents in the land around the hive. We understand our place through these familiar reacquaintances with what we first tasted and became kin with in the nursery.

Opening the Hive: Let Your Heart Precede You

I don't open my hives as often as most beekeepers because I know how important it is to keep the heat and scent within the hive. When I do go into the hive, it's for good reason, like in early fall to confirm that the bees have stored enough honey to make it through the winter, or later in spring to see if there is enough leftover winter honey to harvest before the flow. Sometimes I check the volume of honey simply by lifting the back of the hive an inch and feeling its heft. If you do this a few times through the year, you'll know by feeling the weight if the hive is full or there is not enough extra honey for the colony to make it through the winter.

One way beekeepers can help their bees is simply to open the hive less often. New beekeepers especially seem to have a burning need to get into their hives and see what is going on. While it is an education to peek inside, it is always a setback—sometimes a major one—for the bees.

Bees work hard to create a protective, healing seal in their hive and to keep a specific and precise temperature for the brood. It takes the bees about a day-and-a-half to restore the balance in their hive after it has been opened and inspected. Still, there are times when hives must be opened so that the beekeeper can check on the hive's health, add or remove a hive box, or gather honey.

When I need to open my hives, I am always conscious of where I am standing, what the weather is doing, and how the bees are feeling that day. Standing in the flight path, opening a hive in cold weather and dangerously chilling your bees, or not respecting warning buzzes or bumps from guard bees are recipes for certain disaster.

I open my hives on warm days when the bees let me know they are not dealing with their own stresses that day. I always move very slowly and quietly, which brings calmness to the bees and to me, too. Most important, I bring loving and caring intent into my actions, and I purposely pour love into the open hive. I cannot overstate how important intention and loving-kindness are to the bees. They speak of it often.

IN OUR OWN WORDS

Breaking the propolis seal opens the hive to outer influences that are hard to control.

Some human participation is okay. But this must be done with presence and attention to the sacredness of our hive. Open our home with a ritual, full of prayer and love, as you come into communion with us.

Many beekeepers open a hive with focus on the colony, but with little genuine and heartfelt interaction. We are not simply a science project. When you open our hive respectfully, open yourself, too, to being blessed. Ask to open and wait for the answer. Open and pause. Let your heart precede you. Feel the emanation from the hive and let it enter you. Move slowly and gently. You are in our home, a holy place.

Learn the difference between hovering, observing, and being present. The energy moves between us. Open your heart to feel the presence of our family, our hive, and this expanding sphere of presence between us. At the point where familiarity dwells, let us embrace you, too.

Daily give your blessing to the bees you are partnered with. We do know it, and it strengthens us to be blessed, as it does all beings.

The Whole and Holy Life:
The Evolution of All Beings

Co-evolution thrills me because I see the interrelatedness of life and how we "merry make our way" into the future.

When Joseph and I bought our land, we entered into a marriage-like partnership with the farm. We believe the farm is a cohesive, living entity and far more than the sum of plants, soil, animals, and insects that live in and upon it. Often landowners think of land, plants, trees, and animal life as resources, and within that thought come questions about how we can use those resources, often for profit and not always with the land's best interest at heart. When Joseph and I plan what to do next on our farm, we consider the farm's perspective. Quite often the farm's opinion outweighs our own, and we are bettered by it.

By our fourth year on the farm, we had chickens, honeybees, and a small garden. Then I began dreaming of cows—cows in lush green pastures, milking cows, cows mooing to us—all things I'd never dreamed about before. It was an outright peculiarity, and I took it to mean the farm was asking for cows. Neither of us had been inclined to get a cow, but the farm wanted cows and presented us with many reasons why a cow would be good for us: improving the land and our health, creating a loving relationship with a new animal, helping us develop commitment and deepen friendships within our community.

That seemed like a tall order, but we went ahead and bought a lovely cow, who became mother to many. Over time, we sold or traded the heifers to our neighbors. After a few years, we called all the neighborhood cows together and built a community herd, sharing pastures and chores. I would not have imagined such a thing if we hadn't paid attention to those incongruous thoughts. The harmony and satisfaction that come from living in agreement with our land and farm bring us great joy.

On a personal level, I want to live my life in such a way that the presence of God is all around me, as I imagine it is when these mundane and miraculous moments occur. As such, I desire to witness and support that which is highest in each person I know. I hold that someday I may be a better reflection of God to know the world through me. In the meantime, I seek to build life force around us in every way I can. Our farm is a good place to do this.

Wherever you live is a good place to do this. Somewhere nearby, a bee colony may be tending your home, your breath, your sunlight, flowers, and nature spirits. The bees who give attention to your garden may be from your own hives, or perhaps wild bees in a wooded, high bower have enfolded you into their universe. You are already part of a unity—maybe more than one—and you have a gift to contribute. Step forward and give your love and your joy to this glorious task.

IN OUR OWN WORDS

Each piece of land dwelled upon bespeaks the balance of all the beings present upon that land and of the multiple dimensions of the land's life beings. This is written as a code, a tally of the balance (or unbalance) that place bears. Imagine the code suspended over the land, like a street address, able to be read by anyone cognizant of these forces. Thus it sits, an open-ended equation that can be added to, subtracted from, multiplied with, or separated by division. Certain combinations open new possibilities, acting by their presence as harbingers of possible development.

Humans make a contribution by developing the higher self. The community present on the land is influenced by the emotional tones carried in each human. These powerful singings help or hinder the development and expression of all beings present. Humans who dwell in strongly negative energies are capable of undoing eons of expression and can bring the land presence to doldrums; forward advancement comes to a pause. Such human action may not seem significant on its own, but in party to all the presences, it can stall, influence, or hasten co-evolution.

Nature seeks not only the evolution of all beings in their own turn, but also the co-evolution wherein all are party to each other's blended progress.

In such a way, bees and humankind co-evolve. But while humans may well "keep bees," the kingdom of bees longs to sing the song of shared awareness, of our mutual caring for one another. Bees enrich and harmonize the environment each day, helping Nature in so many ways to fulfill the evolutionary directive.

When humankind becomes present, knowing of these relationships, the code activates. When barren intellect outweighs knowledge of the relationships, we falter in our development and introduce a division; we separate our actions into what benefits us alone. Progress halts because the code is unbalanced, and needed components and energies are deprived of their valences.

To evolve in the current time, love needs to be present. When working with the bees and with each other, ask your hands and your hearts to be gentle; let your mind be guided toward actions that fulfill the purpose of bees in the world, not for shortsighted solutions or harvests that unbalance the colony's needs. Learn to respect the sanctity of the bond. Use that knowledge to become kinder and more compassionate, and walk forward into the future—all of us together—emanating and enveloped in our shared love and awakening.

Epilogue: Bee Sleepover

I hope through reading this book, you are more inclined to percive the world as a bee would and to care for this earth so all may live and flourish.

Alas, most people still are afraid of bees. Many of those people have little or no experience with bees on which to base their fearfulness. Why are so many people scared? The media don't tell us that bees are inclined to be good-natured. News programs rarely distinguish bees from hornets, yellow jackets, and wasps; all are called bees. Until recently, bees didn't generally make news, because bees are very quiet, and beekeeping has generally been a peaceful hobby that many people take to in retirement. Then came alarming stories of "killer bees" aggressively attacking people and animals, and terrifying tales of giant Asian wasps with nasty dispositions.

When I work the bee booth at the county fair, people often timidly ask how to protect themselves from bee attacks. I tell them my experience, how I am around bees most every day and rarely have a moment of concern.

Last spring a friend gave me a swarm she had collected from her wild beehive. It was nearly nightfall when the bees arrived on our farm. I know better than to try and move bees into a new hive toward dark, but I hoped it would be simple and quick. I opened the box and saw thousands of crawling bees, so I tipped the box upside down and shook half the bees into the new hive. What I didn't know was that the other half were clustered on a big branch that was wrapped in a pillowcase inside the box. Once the surface bees were moved, I saw that the bees below were tangled in cloth and branches. Alas, I couldn't turn back. In the dark I tried to shake out the pillowcase, but the bees inside didn't want to release their hold because they were afraid of falling away from their family.

A half hour later, I had almost everyone out of the container, off the branch, and unstuck from the folds of the pillowcase. With all the branch and cloth shaking, a good number of bees had fallen onto the floor of the bee house, where I keep a number of colonies under a permanent roof structure. Bees don't like being on the floor, so every little lost bee moved toward the only tall thing they could find: my legs. They climbed over my sneakers, up my socks, and underneath my jeans. I felt the tickle as each one climbed my legs all the way to the top of my knees, as far as my jeans allowed them to go. Being in the midst of the transfer, I told these little bees not to be afraid and to find a safe place and wait for me. I would tend to them as soon as I could.

Another half hour later, I tucked the new hive in and told the lost bees we were going to wend our way down to the farmhouse where I had light and could find them all. I walked across the field to the house slowly, so my jeans didn't bind anyone. Once in the bathroom, I closed the door to keep us all in one place. Then I rolled down my jeans at about ice-melting speed, finding and carefully lifting off each bee that was on my thighs, shins, and socks.

All in all, I collected fourteen bees from the inside of my pants. I am so proud of these little bees, because they went through an adventure that could have been traumatic and not one of them got upset. Fourteen bees; no stings. Bless the sweetness of each one.

In daylight, the foraging honeybee is an adventurer, an intrepid explorer; but then she hurries home before dusk. When I find a bee in the house, I gently carry her outdoors, reassuring her of her safety and well-being.

I always check the west windows of our south-facing garage late in the day for errant bees. The bees who live in our north hives often forage in the southeast field, and if they take a shortcut back through our garage, with its open door, they get caught against the unopened west windows. Why do they get stuck? Because a honeybee cannot override her sense of direction. To guide her home, each bee has a tiny speck of magnetic oxide nanoparticles concentrated in her antennae and abdomen. When a bee flies up against that west-facing window, she knows absolutely that home is in that direction if only she could get past the window. She will continue buzzing against the window until she exhausts herself and dies. You would imagine these bees would eventually figure out they cannot get through the closed window, and then they'd go back out through the open garage door, but they don't work that way. Curiously, I also have seen butterflies, moths, crane flies, mason bees, and ladybugs in the windows, and they nearly always find their way out quickly. Only the honeybees get caught there and die.

Sometimes after dark, I find a bee or two in the house, buzzing against a light fixture. A bee can't fly at night without the sun to help her navigate home. If I release her outside, she will simply get lost and most likely die. Because we have many hives here, I can't tell which one is her home, so it is impossible to deliver her to the correct hive. I have tried putting an evening bee on the front step of the nearest hive, and sometimes that works—the little girl slips right inside. But just as many times, the guard bees come out and give the stranger a rough once-over. When that happens, I suddenly have a lost and scared bee on my hands. Alas, what to do?

I used to put a dish with a dab of honey on my counter and then place a glass jar over the dish to keep her confined until morning. But I noticed the bee often spent a stressful evening buzzing against the glass, trying to get out. When I thought of her natural environment, I came up with something different: the bee sleepover jar.

Trapped foragers are stressed about being out after dark, so I wanted to re-create a hive-like place, with their natural scents of wax comb, fragrant propolis, sweet honey, and water, to give a sense of familiarity and safety. I placed two small chunks of empty comb and a tiny smidge of highly scented propolis inside a pint glass jar. I pressed the edges of the wax so it adhered to the sides, leaving room between the combs so it appeared similar to the inside of a hive. Using a toothpick, I filled a cell near the top with honey, and I put a single drop of water in another cell—a pinhead-size drop. Then I poked tiny air holes into the metal cap or used the screened cover of a sprouting jar.

Now when I find a bee after dark, I invite her to this bee sleepover jar. I tell her I have a safe place for her to rest until morning, and using my finger, I gently place her inside and let her walk onto the comb. Then I screw on the ventilated top and put the jar and bee in a darkened room. Sometimes I find two or three bees out for the evening, buzzing about in my kitchen. I put them together, and they all have a friendly sleepover in the jar.

Once I have the bees safely settled in for the evening, I write myself a note to remind me in the morning that my little lost bees are in the jar. I write the note because I have twice awakened and gone about my day, forgetting the bee jar until later. By then, their time apart from the colony was too long, and the bees had died. Now I tape notes on the bathroom mirror, my office desk, and the kitchen table saying, "Bees in bathroom!"

After the sun comes up, I peek at the bees in the sleepover jar, and nearly always they are just fine. Once they start moving around, I walk up the path to the bee yard and open the jar. They immediately fly home.

You, too, can create a small bee bed-and-breakfast with a pint jar and some pieces of comb. Each evening at dusk, you can make the rounds on your land and home and check that no bees are caught in the greenhouse, garage, kitchen, or elsewhere. This attentiveness to bees around you—noticing stragglers in your midst—is a simple and wonderful meditative practice on awareness and kindness, benefiting you and the bees.

Does it matter that one little bee makes it through the night?

I believe it does. Being kind to one bee, even when it likely won't make much difference to the hive or even the bee community, is a good thing for us humans to do. Maybe the world won't change because I saved a bee. But, too often, the callousness of my inattention denies me the opportunity to develop benevolence. No being is inconsequential; every life matters. When we treat all beings as deserving of our consideration, even a little bee can assist us in our task of becoming gentler, more thoughtful, more human.

Glossary

Arc of Creation: The spirit-filled channel of birth through which all bees pass as they are being born. It is here that the drones sing their World Song to the new pips, imbuing them with knowledge of their history and future, as the maidens sing of the tasks they do in the present.

Bar: The upper part of a wood frame to which bees adhere their comb. Typically a frame is four-sided, like an empty picture frame, but in top-bar and Warre hives, the bar is a single, freestanding wooden strip with no sides.

Bee bread: Food for baby bees, made from pollen collected in the field, mingled with saliva, and fermented. Also called *field cake*.

Bee space: The distance between parallel combs—a distance wide enough for two bees going in opposite directions to easily pass each other.

Box: See *hive box*.

Building up: A queen's greatly increased egg laying in early spring, meant to increase the population of maidens when most flowers bloom.

Cluster: A number of the bees snuggled up together to conserve heat and reduce energy expenditure in winter. See *torpor*.

Comb: The skeleton of a bee's hive, built in flat, vertical panels using wax from the bees' bodies. Comb is covered with hexagonal cells.

Drone: Male bee.

Emergency queen: A new queen quickly created from a maiden pip when the colony's queen has unexpectedly died. The pip is fed royal jelly, which changes her into a queen capable of laying eggs, thus keeping the hive alive. Once the colony has survived the emergency, the maidens will often create a new queen from a queen egg, to replace the emergency queen. (More can be learned about this in Gunther Hauk's book *Toward Saving the Honeybee*.)

Field cake: See *bee bread*.

Frame: A four-sided structure that holds comb in a Langstroth hive, or is used to repair a hive with broken comb.

Hive box: The wooden container used by beekeepers to contain bee colonies. Langstroth and Warre hives are made of multiple boxes piled vertically atop one another.

Honey: The sweet, nutritious, and flavorful liquid made from the dehydrated nectars that bees gather from plants. Honey is a preservative, and it never goes bad.

Honey bound: What happens when the bees have so much nectar they fill every inch of the comb in their home with more and more honey, until there are no more empty cells for the queen to lay her eggs in. If no honey is removed, the hive's population will begin to reduce in size, to its detriment.

Langstroth hive: The most common hive used around the world. Its benefit is that the frames can be easily moved and manipulated by humans; however, ease of use for humans doesn't mean it's the best home for bees. For this reason, many alternative beekeepers use other types of hive homes. The frames of "Langs" often use premade plastic comb, instead of open space where bees can build their own wax comb.

Larva: The middle stage of a pip, after the first three days as an egg and before it metamorphoses into a pupae.

Lumen: A point where the light emerges from the earth and extends out to the heavens. It is where a virgin queen, as she mates with the drones, comes into understanding of her role and delivers her message to the heavens.

Maiden: Female bee who does all the chores the colony requires to stay in good health, including caring for the pips in the nursery and finding and bringing home nectar, pollen, and the ingredients of propolis. Maidens are called "worker bees" in beekeeping. Maidens' reproductive systems are not functional as long as their hive has a healthy queen.

Nasanov gland: Scent gland that maidens use to put out a scent that helps young forager bees find their way back to the hive or, with swarms, that marks a new home.

Nuptial flight: A young virgin queen's first flight out of the hive to mate with as many drones as she can, thus ensuring plenty of sperm to fuel her egg production for many years.

Pinching the queen: In conventional beekeeping, the process of removing and killing a year-old queen or any queen presumed to be lacking in fertility or having qualities a beekeeper doesn't want in their bees. After a queen is pinched, she is replaced with another, usually one artificially bred and unrelated to the colony. Pinching the queen is tremendously stressful to a colony.

Pip: Baby bee before it hatches.

Pollen: A powdery substance that contains the male reproductive part of a flowering plant. When moved by bees to another flower, the sperm in the traveling pollen fertilizes the female pistil and causes germination.

Pollination: The transfer of pollen, which carries a flower's genetic material, to another flower via bees. A foraging bee visits flowers while gathering nectar. While on a flower, the bee's static electrical charge causes pollen grains to adhere to the bee's body. When the bee alights on the next flower, the male pollen grains carried by that bee mingle with the flower's female parts, ensuring that the flowers are sexually fertilized.

Propolis: A substance bees make by mixing resin from tree buds and plants with wax, essential oils from plants, and other materials. Bees use propolis to strengthen every surface of the hive. It seals the hive so no undesirable elements or drafts can enter. It is stored throughout the hive and is used to keep the comb adhered to the bar. Bees also inhale the vapors from propolis as medicine, taking advantage of its antibiotic, antifungal, antioxidant, and antiviral properties.

Queen: The sole fertile female in a hive, responsible for laying eggs to provide the ongoing renewal of the hive's population.

Queen excluder: A barrier used in Langstroth hives to keep the queen from getting into the honey supers and laying eggs. Using a queen excluder is not necessary, as the queen knows eggs belong in the nursery.

Queenless: A colony in which a queen has died and the bees are grieving her loss. The bees in a queenless colony often make a moaning sound to express their grief.

Queenright: A colony in which the queen is healthy and productive.

Queen substance: A pheromone exuded from the queen's mandibular glands that produces a scent unique to that queen. The scent keeps the colony unified and joy-filled in its tasks and suppresses reproductive hormones in the maidens, keeping them infertile.

Royal jelly: A substance that madien and drone bees produce from their hypopharyngeal glands and feed to pips during their first few days of gestation. Queen pips, however, receive a steady diet of royal jelly throughout their gestation. An ascended queen continues being fed a diet of royal jelly during her time as queen. Royal jelly provides some immunity against bacterial infections and thus helps protect the bees right from the start of their lives.

Scarp: A highly intentioned focal point within the lumen; the place where the drones mate with virgin queens. Humans call these places "drone congregation areas." They are located a few hundred feet above the ground.

Seed: The incipient aspect of sperm or egg; that which will become a bee.

Supers: Individual boxes stacked up to form a standard vertical hive within a Langstroth hive. Supers are named according to their size or function. Big to small, they are called (1) deeps, used on the bottom as brood boxes; (2) mediums, which are used in the middle and may be used for brood or honey; and (3) small supers, a topmost box used only for honey.

Swarm: The approximate two-thirds of a healthy colony that departs a full hive to form a new colony in a new hive home. A swarm contains the colony's queen and a population of mature bees.

Swarming: The process in which most of the bees in a healthy colony leave their old home and, flying together in a tumultuous, ecstatic cloud, move to establish a new colony in a new location.

Top-bar hive: Horizontal hives of a fixed size. Unlike hives with four-sided frames, top-bar hives have only a single bar at the top, from which bees construct their comb as they desire. A colony living in a top-bar hive should be managed minimally to keep the fixed-size box from becoming honey bound. One advantage of a top-bar hive is that the beekeeper doesn't need to lift heavy boxes—only single frames—so maneuvering a top-bar hive doesn't require physical strength.

Torpor: In winter, during very cold weather, bees may go into torpor, a state of semi-hibernation that helps them conserve their energy and food resources. Bees in torpor may look quite dead. It may take as long as three days for them to rouse into wakefulness once they are warmed.

Warre hive: A vertical, top-bar hive. It was invented by Abbé Warré, a French monk who, after many decades of observing natural bee behavior, determined that the most important qualities of a human-made hive are the preservation of scent and temperature. Its novel roof system keeps moisture or mold from harming the bees, and its inviolability lets the bees design the interior according to their needs. Honey is harvested by the beekeeper taking off the topmost box and sliding an empty box underneath the bottommost box.

Works Cited

Epigraph

"When we step into the world of *Apis mellifera* . . ." Michael Joshin Thiele, "Lessons from the Bees," *Biodynamics* (Fall 2013).

Chapter I

"For just one example of how bees benefit crops . . ." S. Alan Walters and Bradley H. Taylor, "Effects of Honey Bee Pollination in Pumpkin Fruit and Seed Yield," *Horticultural Science* 40, no. 2 (April 2006): 370–373. Available at hortsci.ashspublications.org/content/41/2/370.full.pdf.

Chapter II

"Truly, the term 'labor of love' would apply to the workers . . ." Gunther Hauk, *Toward Saving the Honeybee* (Junction City, OR: Biodynamic Farming and Gardening Association, 2002): 32.

"Conventional beekeepers see the introduction of a queen bee into a colony . . ." Erik Berrevoets, *Wisdom of the Bees: Principles for Biodynamic Beekeeping* (Great Barrington, MA: Steiner Books, 2009): 95.

Chapter III

"Phil Chandler, author of *The Barefoot Beekeeper* . . ." Phil Chandler, The Barefoot Beekeeper, biobees.com.

"Some of its properties defy not just chemical analysis . . ." and "The beehive is a symbol of how simpler substances . . ." James Fearnley, *Bee Propolis: Natural Healing from the Hive* (London: Souvenir Press, 2001): 4, 37.

"Silent robbing occurs when robbing has reached a state . . ." Roger Morse and Ted Hooper, *The Illustrated Encyclopaedia of Beekeeping* (Poole, Dorset, UK: Blandford Press, 1985): 343.

Chapter IV

"Because of the way that bee bread is inoculated, matured, and distributed, . . ." Heather R. Mattila, Daniela Rios, Victoria E. Walker-Sperling, Guus Roeselers, and Irene L. G. Newton. "Characterization of the Active Microbiotas Associated with Honey Bees Reveals Healthier and Broader Communities when Colonies Are Genetically Diverse," *PLOS ONE* 7, no. 3 (March 12, 2012): e32962. Available at journals.plos.org/plosone/article?id=10.1371/journal.pone.0032962.

Chapter V

"The single most serious factor . . ." Gunther Hauk, *Toward Saving the Honeybee* (Junction City, OR: Biodynamic Farming and Gardening Association, 2002): 30.

Chapter VI

"Beekeeping advances civilization because it makes man strong. . . . ," "Like bees, humans need nourishment that carries over into our bodies . . . ," and "The consciousness of a beehive, not the individual bees, is of a very high nature . . ." Rudolf Steiner, "Nine Lectures on Bees" (lectures, Dornach, Switzerland, February, 1923). Transcripts, translated by Marna Pease and Carl Alexander Mier, are available at biobees.com/library/biodynamics/SteinerBeeLectures.pdf.

Resources

Following is a list of books, DVDs, websites, and articles to deepen your knowledge and appreciation of the honeybee.

Books

Bush, Michael. *The Practical Beekeeper: Beekeeping Naturally.* X-Star Publishing, 2011.

Fearnley, James. *Bee Propolis: Natural Healing from the Hive.* London: Souvenir Press, 2001.

Frey, Kate, and Gretchen LeBuhn. *The Bee-Friendly Garden.* Berkeley, CA: Ten Speed Press, 2016.

Hauk, Gunther. *Toward Saving the Honeybee.* Junction City, OR: Biodynamic Farming and Gardening Association, 2008.

Heaf, David. *The Bee-Friendly Beekeeper: A Sustainable Approach.* West Yorkshire, UK: Northern Bee Books, 2010.

Herboldsheimer, Laurie and Dean Stiglitz. *The Complete Idiot's Guide to Beekeeping.* Indianapolis, IN: Alpha Press, 2010.

Kornberger, Horst. *The Global Hive.* Hamilton Hills, Australia: School of Integral Arts Press, 2012.

Lissantheia, Aimee, and Tim Lukowiak. *The Amazing Adventures of Melissa Bee.* Rolling Hills Estates, CA: Beloved Press, 2015.

Morse, Roger A., and Ted Hooper. *Illustrated Encyclopaedia of Beekeeping.* Poole, Dorset, UK: Blandford Press, 1985.

Siegel, Taggart, and Jon Betz, eds. *Queen of the Sun: What Are the Bees Telling Us?* West Sussex, UK: Clairview Books, 2011.

Tautz, Jurgen. *The Buzz About Bees: Biology of a Superorganism.* Berlin and Heidelberg, Germany: Springer Verlag, 2008.

Weiler, Michael. *Bees and Honey: From Flower to Jar.* Trowbridge, UK: Cromwell Press, 2006.

The Xerces Society, *Attracting Native Pollinators.* North Adams, MA: Storey Publishing, 2011.

DVDs and Videos

Alternative Beekeeping Using the Top Bar Hive and the Bee Guardian Methods, Corwin Bell, available at BackyardHive (BackYardHive.com). Every beekeeper should have this video.

Queen of the Sun, award-winning and visually stunning documentary directed by Taggart Siegel about bees and beekeepers who are working in bee-centric ways.

Videos by Michael Joshin Thiele. Thiele always engages bees with respect and deep appreciation. Visit his website, Gaia Bees (GaiaBees.com), to see his latest videos.

Jacqueline's website, Spirit Bee (SpiritBee.com), offers movies, audio files, and images of colonies singing the Song of Increase and reflecting other fascinating states of being in the hive, including a good audio file to listen to during a bee-medicine session. Sound by Robin Wise.

Bee Stuff

Mickelberry Gardens (mickelberrygardens.com), organic and treatment-free gifts from the hive, including propolis for healing. (See chapter VI, The Song of Abundance.)

Helpful Organizations and Internet Groups

The Xerces Society for Invertebrate Conservation (xerces.org). Please support this organization. They do vital work to protect bees and other pollinators.

The Organic Beekeepers Group (organicbeekeepers-subscribe@ yahoogroups.com), a treatment-free Internet group that meets annually for a wide-ranging educational conference.

Warre Beekeeping (uk.groups.yahoo.com/neo/groups/ warrebeekeeping/info), David Heaf and other helpful beekeepers who use Warre hives.

Articles

Biodynamics: Agriculture in Service of the Earth and Humanity, Biodynamic Farming & Gardening Association, USA, #281, Fall 2013. "Lessons from the Bees" contains a series of articles by various authors.

Star and Furrow: Journal of the Biodynamic Association (UK), issue no. 120, December 2013. "The Wisdom of Bees" contains a series of articles by various authors.

Adam Frank, "Quantum Honeybees," *Discover,* November 1, 1997, discovermagazine.com/1997/nov/quantumhoneybees1263. Describes mathematician Barbara Shipman's discoveries about the quantum aspects of the bee waggle dance.

Michael Joshin Thiele, "Honoring the Bien," *Lilipoh,* 52, no. 13 (Summer 2008), Lilipoh.com, "Past Issues." Many other fine articles on relating to honeybees in a more sensitive manner can also be found on the *Lilipoh* website.

Websites

Spirit Bee (spiritbee.com) is where Jacqueline Freeman, Susan Chernak McElroy, and Robin Wise post their work with bees, including videos, sound files, books, and photos; further writings about and from the bees; and lovely, hand-designed items that use their bee imagery. Sign up for Jacqueline's bee newsletter. The audiobook *Song of Increase* is available here, too.

BackYardHive (BackYardHive.com) contains information and hive technologies that encourage and enable backyard beekeepers to be successful, founded by Corwin Bell.

The Barefoot Beekeeper (biobees.com) offers information about practical, balanced, treatment-free beekeeping in top-bar hives.

Gaia Bees (GaiaBees.com), founded by Michael Joshin Thiele.

Holy Bee Press (HolyBeePress.com), a crossroads of honeybee conversation and bee salon, founded by Deborah Roberts.

Natural Beekeeping Trust (NaturalBeekeepingTrust.org) is a nonprofit that focuses exquisite attention on the needs of bees.

College of the Melissae: Center for Sacred Beekeeping (CollegeoftheMelissae.com), Laura Bee Ferguson, director.

Michael Bush, the Practical Beekeper (BushFarms.com/bees), has extensive articles on every aspect of healthy, treatment-free beekeeping.

Acknowledgments

I thank the bees, who have been incredibly generous to share this information with me. I hope I have conveyed their brilliant knowledge in a way that helps others come to love them, too. I am grateful to our farm, Friendly Haven Rise, for the wide bounty of beauty and the feeling of love that emanates from this place.

As I wrote this book, I was surrounded by intelligent, devoted people who helped me carry my vision. I thank my husband, Joseph, for creating the marvelous three-dimensional framework that vividly portrayed how all aspects of bee life connect with each other and provided the foundation for this book to light upon. Also, I am grateful that he transcribed many of these teachings as I spoke with the bees, which allowed me to go deeper.

I thank Patti Pitcher for her early and persistent encouragement and her wisdom in all things. I thank Sara Cooper for her insight, support, and delightful friendship. I thank Susan Chernak McElroy for unraveling the structure of this book and making it easy for me to understand, and for coming out and staying at the farm as I wrote so I could ask a hundred questions. Her precious guidance and extensive experience as an author turned hundreds of stray pages into a book. I will always be grateful she put her next book aside and offered to help me with mine. I look forward to more collaboration in the future.

I thank Robin Wise for accompanying me on the bee journey with her photos and audio files to help us see and hear all that surrounds us, and for enriching my life with her astute observations and commitment to holding bees in a sacred manner. Her eminent skills brought forth *Song of Increase* as an audiobook. Glenna Rose helped with editing; I felt like a better person when she finished.

I thank my super-smart father, Charlie Entwistle, for his inspiration and advice that I should do what I love and figure out ways to make those things become my work. His generosity and love make my heart happy. I am overjoyed he got to read this book before he passed. My dad was not particularly fond of bees, but after he read the book, one day I watched him gently pick up a bee on his finger and commune with it. I hope this book helps a few billion more folks make that shift. My stepmom, June, planted the seed by publishing her kumquat cookbook (*AJ's Place: Home of the Kumquat Queen*) a decade ago, and ever since, I have wondered how I could do that.

I thank these people whose presence in the world I thoroughly enjoy: fiery Amber Ham; generous Catherine Miller-Smith; my sweet mother, Jesse, who gifted me with intuition; compassionate Brenda Wilson; caring and diligent Bonnie York; perceptive Bambi Dore; my brother, C.W. (the first person to introduce me to bees); benevolent Deborah Entwistle; Kay Gleason and her marvelous sense of humor; ever-so-kind Nathan Rausch; my beloved Aunt Ruth; Corwin Bell, who sees beyond; radical agricultural wizard Steve Storch; full-of-love Debra Roberts; depthful Michael Joshin Thiele; always persistent Dee Lusby; creative Brenda Calvert; loving B.J. Schulte; good friend Susan Tripp; soulful Linda Bieneufeld; sensitive and wise John Takacs; Thomas Chavez, who introduced me to grace states; diligent Wes Burch; bright Summer Michaelson; ground-shaker Paul Wheaton; poet and farmer Rick Sievers; David Heaf, for his learned science review and the many discussions that sprung from that; and these lovely women who have been my best friends at different times in my life: Jan Swindell, Margot Datz, and Susan Belfiore.

This book was self-published in 2015 and had a rocking good start. My new publisher, Sounds True, helped make the book even better. Additional applause to my agent, Joanne Wyckoff, and Amy Rost of Sounds True, who walked me through every word. Special thanks to Hedgebrook, the writing retreat that offered me the gift of quiet time and a place to put this book together.

About the Author

Jacqueline Freeman is a relational beekeeper, a biodynamic farmer, and a pioneer in the field of natural beekeeping. She is gifted in perceiving nature intelligences, particularly that of honeybees, and has spent many years working alongside them with an open and prayerful heart.

Jacqueline teaches insightful bee classes at her farm and honeybee sanctuary. The documentary movie and book *Queen of the Sun* showed her caring work as a gentle swarm rescuer. Her bee articles appear in national magazines, and she's a featured speaker at national conferences for organic and treatment-free beekeepers, permaculture, and sustainable agricultural events. She has worked in the Dominican Republic with rural beekeepers, helping them create healthy bees through respectful and treatment-free beekeeping. Annually she hosts a multi-day conference for radical beekeepers.

Jacqueline's website, Spirit Bee (SpiritBee.com), lists events and classes and has beautiful photos, movies, audio files, and hive gifts, many created with Robin Wise. You are invited to read more of her bee communications and see videos of her working with her bees throughout the year.

Jacqueline and her husband, Joseph, live on their farm in southwest Washington State.

About Sounds True

Sounds True is a multimedia publisher whose mission is to inspire and support personal transformation and spiritual awakening. Founded in 1985 and located in Boulder, Colorado, we work with many of the leading spiritual teachers, thinkers, healers, and visionary artists of our time. We strive with every title to preserve the essential "living wisdom" of the author or artist. It is our goal to create products that not only provide information to a reader or listener, but that also embody the quality of a wisdom transmission.

For those seeking genuine transformation, Sounds True is your trusted partner. At SoundsTrue.com you will find a wealth of free resources to support your journey, including exclusive weekly audio interviews, free downloads, interactive learning tools, and other special savings on all our titles.

To learn more, please visit SoundsTrue.com/freegifts or call us toll-free at 800.333.9185.

SOUNDS TRUE
many voices, one journey